AGRICULTURE COURSE

The Birth of the Biodynamic Method

Eight lectures given in Koberwitz, Silesia, between 7 and 16 June 1924

RUDOLF STEINER

RUDOLF STEINER PRESS

Translated by George Adams

Rudolf Steiner Press
Hillside House, The Square
Forest Row, RH18 5ES

www.rudolfsteinerpress.com

Published by Rudolf Steiner Press 2004. Reprinted 2008, 2011, 2012, 2016

First published in English in 1958

Originally published in German under the title *Geisteswissenschaftliche Grundlagen zum Gedeihen der Landwirtschaft* (volume 327 in the *Rudolf Steiner Gesamtausgabe* or Collected Works) by Rudolf Steiner Verlag, Dornach. This translation is published by permission of the Rudolf Steiner Nachlassverwaltung, Dornach

A catalogue record for this book is available from the British Library

ISBN 978 1 85584 148 2

Cover by Andrew Morgan Design
Printed and bound by Gutenberg Press, Malta

FSC www.fsc.org MIX Paper from responsible sources FSC® C022612

The paper used for this book is FSC-certified and totally chlorine-free. FSC (the Forest Stewardship Council) is an international network to promote responsible management of the world's forests.

CONTENTS

PREFACE

BY

EHRENFRIED PFEIFFER, M.D.(Hon.)

In 1922/23 Ernst Stegemann and a group of other farmers went to ask Rudolf Steiner's advice about the increasing degeneration they had noticed in seed-strains and in many cultivated plants. What can be done to check this decline and to improve the quality of seed and nutrition? That was their question.

They brought to his attention such salient facts as the following: Crops of lucerne used commonly to be grown in the same field for as many as thirty years on end. The thirty years dwindled to nine, then to seven. Then the day came when it was considered quite an achievement to keep this crop growing in the same spot for even four or five years. Farmers used to be able to seed new crops year after year from their own rye, wheat, oats and barley. Now they were finding that they had to resort to new strains of seed every few years. New strains were being produced in bewildering profusion, only to disappear from the scene again in short order.

A second group went to Dr. Steiner in concern at the increase in animal diseases, with problems of sterility and the widespread foot-and-mouth disease high on the list. Among those in this group were the veterinarian Dr. Joseph Werr, the physician Dr. Eugen Kolisko, and members of the staff of the newly established Weleda, the pharmaceutical manufacturing enterprise.

Count Carl von Keyserlingk brought problems from still another quarter. Then Dr. Wachsmuth and the present writer went to Dr. Steiner with questions dealing particularly with the etheric nature of plants, and with formative forces in general. In reply to a question about plant diseases, Dr. Steiner told the writer that plants themselves could never be diseased in a primary sense, "since they are the products of a healthy etheric world." They suffer rather from diseased conditions in their environment, especially in the soil; the causes of so-called plant diseases should be sought there. Ernst Stegemann was given special indications as to the point of view from which a farmer could approach his task, and was shown some first steps in the breeding of new plant types as a first impetus towards the subsequent establishment of the biological-dynamic movement.

In 1923 Rudolf Steiner described for the first time how to make the bio-dynamic compost preparations, simply giving the recipe without any sort of explanation—just "do this and then that." Dr. Wachsmuth and I then proceeded to make the first batch of preparation 500. This was then buried in the garden of the "Sonnenhof" in Arlesheim, Switzerland. The momentous day came in the early summer of 1924 when this first lot of 500 was dug up again in the presence of Dr. Steiner, Dr. Wegman, Dr. Wachsmuth, a few other co-workers and myself. It was a sunny afternoon. We began digging at the spot where memory, aided by a few landmarks, prompted us to search. We dug on and on. The reader will understand that a good deal more sweating was done over the waste of Dr. Steiner's time than over the strenuousness of the labour. Finally he became impatient and turned to leave for a five o'clock appointment at his studio. The spade grated on the first cowhorn in the very nick of time.

Dr. Steiner turned back, called for a pail of water, and proceeded to show us how to apportion the horn's contents to the water, and the correct way of stirring it. As the author's walking-stick was the only stirring implement at hand, it was pressed into service. Rudolf Steiner was particularly concerned with demonstrating the energetic stirring, the forming of a funnel or crater, and the rapid changing of direction to make a whirlpool. Nothing was said about the possibility of stirring with the hand or with a birch-whisk. Brief directions followed as to how the preparation was to be sprayed when the stirring was finished. Dr. Steiner then indicated with a motion of his hand over the garden how large an area the available spray would cover. Such was the momentous occasion marking the birth-hour of a world-wide agricultural movement.

What impressed me at the time, and still gives one much to think about, was how these step-by-step developments illustrate Dr. Steiner's practical way of working. He never proceeded from preconceived abstract dogma, but always dealt with the concrete given facts of the situation. There was such germinal potency in his indications that a few sentences or a short paragraph often sufficed to create the foundation for a farmer's or scientist's whole life-work; the agricultural course is full of such instances. A study of his indications can therefore scarcely be thorough enough. One does not have to try to puzzle them out, but can simply follow them to the letter.

Dr. Steiner once said, with an understanding smile, in another, very grave situation, that there were two types of people engaged in anthroposophical work: the older ones, who understood everything, but did nothing with it, and the younger ones, who understood only partially or not at all, but immediately put suggestions into practice. We obviously trod the younger path in the agricultural movement, which did all its learning in the hard school of experience. Only now does the total picture of the new impulse given by Rudolf Steiner to agriculture stand clearly before us, even though we still have far to go to exhaust all its possibilities. Accomplishments to date are merely the first step. Every day brings new experience and opens new perspectives.

* * *

Shortly before 1924, Count Keyserlingk set to work in dead earnest to persuade Dr. Steiner to give an agricultural course. As

Dr. Steiner was already overwhelmed with work, tours and lectures, he put off his decision from week to week. The undaunted Count then dispatched his nephew to Dornach, with orders to camp on Dr. Steiner's doorstep and refuse to leave without a definite commitment for the course. This was finally given.

The agricultural course was held from June 7 to 16, 1924, in the hospitable home of Count and Countess Keyserlingk at Koberwitz, near Breslau. It was followed by further consultations and lectures in Breslau, among them the famous "Address to Youth." I myself had to forgo attendance at the course, as Dr. Steiner had asked me to stay at home to help take care of someone who was seriously ill. "I'll write and tell you what goes on at the course," Dr. Steiner said by way of solace. He never did get round to writing, no doubt because of the heavy demands on him; this was understood and regretfully accepted. On his return to Dornach, however, there was an opportunity for discussing the general situation. When I asked him whether the new methods should be started on an experimental basis, he replied: "The most important thing is to make the benefits of our agricultural preparations available to the largest possible areas over the entire earth, so that the earth may be healed and the nutritive quality of its produce improved in every respect. That should be our first objcctive. The experiments can come later." He obviously thought that the proposed methods should be applied at once.

This can be understood against the background of a conversation I had with Dr. Steiner *en route* from Stuttgart to Dornach shortly before the agricultural course was given. He had been speaking of the need for a deepening of esoteric life, and in this connection mentioned certain faults typically found in spiritual movements. I then asked, "How can it happen that the spiritual impulse, and especially the inner schooling, for which you are constantly providing stimulus and guidance bear so little fruit? Why do the people concerned give so little evidence of spiritual experience, in spite of all their efforts? Why, worst of all, is the will for action, for the carrying out of these spiritual impulses, so weak?" I was particularly anxious to get an answer to the question as to how one could build a bridge to active participation and the carrying out of spiritual intentions without being pulled off the right path by personal ambition, illusions and petty jealousies; for these were the negative qualities Rudolf Steiner had named as the main inner hindrances. Then came the thought-provoking and surprising answer: "This is a problem of nutrition. Nutrition as it is to-day does not supply the strength necessary for manifesting the spirit in physical life. A bridge can no longer be built from thinking to will and action. Food plants no longer contain the forces people need for this."

A nutritional problem which, if solved, would enable the spirit to become manifest and realise itself in human beings! With this as a background, one can understand why Dr. Steiner said that "the benefits of the bio-dynamic compost preparations should be made available as quickly as possible to the largest possible areas of the entire earth, for the earth's healing."

This puts the Koberwitz agricultural course in proper perspective as an introduction to understanding spiritual, cosmic forces and making them effective again in the plant world.

In discussing ways and means of propagating the methods, Dr. Steiner said also that the good effects of the preparations and of the whole method itself were "for everybody, for all farmers"—in other words, not intended to be the special privilege of a small, select group. This needs to be the more emphasised in view of the fact that admission to the course was limited to farmers, gardeners and scientists who had both practical experience and a spiritual-scientific, anthroposophical background. The latter is essential to understanding and evaluating what Rudolf Steiner set forth, but the bio-dynamic method can be applied by any farmer. It is important to point this out, for later on many people came to believe that only anthroposophists can practise the bio-dynamic method. On the other hand, it is certainly true that a grasp of bio-dynamic practices gradually opens up a wholly new perspective on the world, and that the practitioner acquires and applies a kind of judgment in dealing with biological—*i.e.* living—processes and facts which is different from that of a more materialistic chemical farmer; he follows nature's dynamic play of forces with a greater degree of interest and awareness. But it is also true that there is a considerable difference between mere application of the method and creative participation in the work. From the first, actual practice has been closely bound up with the work of the spiritual centre of the movement, the Natural Science Section of the Goetheanum at Dornach. This was to be the source, the creative, fructifying spiritual element; while the practical workers brought back their results and their questions.

The name, "Bio-Dynamic Agricultural Method," did not originate with Dr. Steiner, but with the experimental circle concerned with the practical application of the new direction of thought.

In the Agricultural Course, which was attended by some sixty persons, Rudolf Steiner set forth the basic new way of thinking about the relationship of earth and soil to the formative forces of the etheric, astral and ego activity of nature. He pointed out particularly how the health of soil, plants and animals depends upon bringing nature into connection again with the cosmic creative, shaping forces. The practical method he gave for treating soil, manure and compost, and especially for making the bio-dynamic compost preparations, was intended above all to serve the purpose of reanimating the natural forces which in nature and in modern agriculture were on the wane. "This must be achieved in actual practice," Rudolf Steiner told me. He showed how much it meant to him to have the School of Spiritual Science going hand in hand with real-life practicality when he spoke on another occasion of wanting to have teachers at the School alternate a few years of teaching (three years was the period mentioned) with a subsequent period of three years spent in work outside, so that by this alternation they would never get out of touch with the conditions and challenges of real life.

The circle of those who had been inspired by the agricultural course and were now working both practically and scientifically at this task kept on growing; one thinks at once of Guenther Wachsmuth, Count Keyserlingk, Ernst Stegemann, Erhard Bartsch, Franz Dreidax, Immanuel Vögele, M. K. Schwarz, Nikolaus Remer, Franz Rulni, Ernst Jakobi, Otto Eckstein, Hans Heinze, and of many others who came into the movement with the passing of time, including Dr. Werr, the first veterinarian. The bio-dynamic movement developed out of the co-operation of practical workers with the Natural Science Section of the Goetheanum. Before long it had spread to Austria, Switzerland, Italy, England, France, the north-European countries and the United States. To-day no part of the world is without active collaborators in this enterprise.

* * *

The bio-dynamic school of thought and a chemically-minded agricultural thinking confronted one another from opposite points of the compass at the time the agricultural course was held. The latter school is based essentially on the views of Justus von Liebig. It attributes the fact that plants take up substances from the soil solely to the so-called "nutrient-need" of the plant. The one-sided chemical fertiliser theory that thinks of plant needs in terms of nitrogen-phosphates-potassium-calcium, originated in this view, and the theory still dominates orthodox scientific agricultural thinking to-day. But it does Liebig an injustice. He himself expressed doubt as to whether the "N-P-K" theory should be applied to all soils. Deficiency symptoms were more apparent in soils poor in humus than in those amply supplied with it. The following quotation makes one suspect that Liebig was by no means the hardened materialist that his followers make him out to be. He wrote: "Inorganic forces breed only inorganic substances. Through a higher force at work in living bodies, of which inorganic forces are merely the servants, substances come into being which are endowed with vital qualities and totally different from the crystal." And further: "The cosmic conditions necessary for the existence of plants are the warmth and light of the sun." Rudolf Steiner gave the key to these "higher forces at work in living bodies and to these cosmic conditions." He solved Liebig's problem by refusing to stop short at the purely material aspects of plant-life. He went on, with characteristic spiritual courage and a complete lack of bias, to take the next step.

And now an interesting situation developed. Devotees of the purely materialistic school of thought, who once felt impelled to reject the progressive thinking advanced by Rudolf Steiner, have been forced by facts brought to light during research into soil-biology to go at least one step further. Facts recognised as early as 1924-34 in bio-dynamic circles—the significance of soil-life, the earth as a living organism, the role played by humus, the necessity of maintaining humus under all circumstances, and of building it up where it is lacking—all this has become common knowledge. Recognition of biological, organic laws has now been added to the earlier realisation of the undeniable dependence of plants upon soil nutrient-substances. It is not too much to say that the biological aspect of the bio-dynamic method is now generally accepted; the goal has perhaps even been overshot. But, important as are the biological factors governing plant inter-relationships, soil structure, biological pest-control, and the progress made in understanding the importance of humus, the whole question of energy-sources and formative forces—in other words, cosmic aspects of plant-life—remains unanswered. The *biological* way of thinking has been adopted, but with a materialistic bias, whereas an understanding of the *dynamic* side, made possible by Rudolf Steiner's pioneering indications, is still largely absent.

Since 1924 numerous scientific publications that might be regarded as a first groping in this direction have appeared. We refer to studies of growth-regulating factors, the so-called growth-inducers, enzymes, hormones, vitamins, trace elements and bio-catalysts. But this groping remains in the material realm. Science has progressed to the point where material effects produced by dilutions as high as 1 : 1 million, or even 1 : 100 million, no longer belong to the realm of the fantastic and incredible. They do not meet with the unbelieving smile that greeted rules for applying the bio-dynamic compost preparations, for these—with dilutions ranging from 1 : 10 to 1 : 100 million—are quite conceivable at the present stage of scientific thinking. Exploration of the process of photo-synthesis—*i.e.* of the building of substance in the cells of living plants—has opened up problems of the influence of energy (of the sun, of light, of warmth and of the moon); in other words, problems of the transformation of cosmic sources of energy into chemical-material conditions and energies.

In this connection we quote from the book *Principles of Agriculture*,* written in 1952 by W. R. Williams, Member of the Academy of Sciences, U.S.S.R.: "The task of agriculture is to transform kinetic solar energy, the energy of light, into the potential energy stored in human food. The light of the sun is the basic raw material of agricultural industry." And further: "Light and warmth are

---

* Translated from the Russian by G. V. Jacks, Director of the Commonwealth Bureau of Soil Science (London, 1952).

the essential conditions for plant life, and consequently also for agriculture. Light is the raw material from which agricultural products are made, and warmth is the force which drives the machinery—the green plant. The provision of both raw material and energy must be maintained. The dynamic energy of the sun's rays is transformed by green plants into potential energy in the material form of organic matter. Thus our first concrete task is the continuous creation of organic matter, storing up the potential energy of human life." And still further: "We can divide the four fundamental factors into two groups, according to their source: light and heat are cosmic factors, water and plant food terrestrial factors. The former group originates in interplanetary space. . ."

Or again: "The cosmic factors—light and heat—act directly on the plant, whereas the terrestrial factors act only through an intermediary (substance)."

We see that the author of this work rates knowledge of the interworking of cosmic and terrestrial factors as the first objective of agricultural science, while ranking organic substance (humus) second on the list of objectives of agricultural production. This is what was published in 1952. In 1924 Rudolf Steiner pointed out the necessity of consciously restoring cosmic forces to growth processes by both direct and indirect means, thereby freeing the present conception of plant nature from a material, purely terrestrial isolation; only through such restoration would it be possible to re-energise those healthful and constructive forces capable of halting degeneration. He said to me, "Spiritual scientific knowledge must have found its way into practical life by the middle of the century if untold damage to the health of man and nature is to be avoided."

* * *

Our research work began with the attempt to find reagents to the etheric forces and to discover ways of demonstrating their existence. Suggestions were given which could only later be brought to realisation in the writer's crystallisation method. Then it was our intention to proceed to expose the weak points in the materialistic conception and to refute its findings by means of its own experimental methods. This meant applying exact analytical methods in experimentation with physical substances, and even developing them to a finer point. We proposed to work quantitatively as well as qualitatively. During my own years at the university, for example, it was my regular practice to lay my proposed course of studies for the new term before Rudolf Steiner for guidance in the choice of subjects. On one occasion he urged me to take simultaneously two—no, three—main subjects, chemistry, physics and botany, each requiring six hours a day. To the objection that there were not hours enough in the day for this, he replied simply, "Oh, you'll manage it somehow."

Again and again, he steered things in the direction of practical activity and laboratory work, away from the merely theoretical.

Suggestions of this kind were constantly in my mind during the decades of work which arose from them. They led me not only to work in laboratories, but also to apply the fundamentals of this new outlook to the management of agricultural projects, both in a bio-dynamic and in an economic sense. Dr. Steiner had insisted on my taking courses and attending lectures in political economy as well as in science, saying, "One must work in a businesslike, profit-making way, or it won't come off." Economics, commercial history, industrial science, even mass-psychology and other such subjects were proposed for study, and when the courses were completed, Dr. Steiner always wanted a report on them. On these occasions he not only showed astounding proficiency in the various special fields, but—what was more surprising—he seemed quite familiar with the methods and characteristics of the various professors. He would say, for example, "Professor X is an extremely brilliant man, with wide-ranging ideas, but he is weak in detailed knowledge. Professor Z is a silver-tongued orator of real elegance. You needn't believe everything he says, but you must get a thorough grasp of his method of presentation."

From these and many other suggestions it was clear what had to be done to promote the bio-dynamic method. There was the big group of practising farmers, whose task it was to carry out the method in their farming enterprises, to discover the most favourable use of the preparations, to determine what crop rotations build up rather than deplete humus, to develop the best methods of plant and animal breeding. It took years to translate the basic ideas into actual practice. All this had to be tried out in the hard school of experience, until the complete picture of a teachable and learnable method, which any farmer could profitably use, was finally evolved. Problems of soil treatment, crop rotation, manure and compost handling, time-considerations in the proper care and breeding of cattle, fruit-tree management and many other matters could be worked out only in practice through the years.

Then there was the problem of coming to grips with agricultural science. Laboratories and field experiments had to provide facts and observational material. I was now able to profit from the technical and quantitative-chemical education urged upon me by Dr. Steiner. This was the sphere in which the shortcomings and weaknesses of the chemical soil-and-nutrient theory showed up most clearly, and where to-day—after more than thirty years—one can see possibilities of building a bridge between recognition of the existence of cosmic forces and exact science.

The first possibility of breaking through the hardened layer of current orthodox opinion came through discoveries that cluster around the concept of the so-called trace elements. Dr. Steiner had pointed out as early as 1924 the existence of these finely dispersed material elements in the atmosphere and elsewhere, and had stressed the importance of their contribution to healthy plant development. But it still remained an open question whether they were absorbed from the soil by roots or from the atmosphere by leaves and other plant organs. In the early thirties, spectrum analysis showed that almost all the trace elements are present in the atmosphere in a proportion of 10^{-6} to 10^{-9}. The fact that trace-elements can be absorbed from the air was established in experiments with *Tillandsia usneodis.* It is now common practice in California and Florida to supply zinc and other trace elements, not *via* the roots, but by spraying the foliage, since leaves absorb these trace elements even more efficiently.

It was found that one-sided mineral fertilising lowers the trace-element content of soil and plants, and—most significantly—that to supply trace-elements by no means assures their absorption by plants. The presence (or absence) of zinc in a dilution of 1 : 100 million decides absolutely whether an orange tree will bear healthy fruit. But in the period from 1924-1930 the bio-dynamic preparations were ridiculed "because plants cannot possibly be influenced by high dilutions."

Zinc is singled out for mention here not only because treatment with very high dilutions of this trace element is especially essential for both the health and the yield of many plants, but also because it is an element particularly abundant in mushrooms. A comment by Rudolf Steiner indicates an interesting connection which can be fully understood only in the light of the most recent research. We read in the Agricultural Course: ". . . Harmful parasites always consort with growths of the mushroom type, . . . causing certain plant diseases and doing other still worse forms of damage. . . . One should see to it that meadows are infested with fungi. Then one can have the interesting experience of finding that where there is even a small mushroom-infested meadow near a farm, the fungi, owing to their kinship with the bacteria and other parasites, keep them away from the farm. It is often possible, by infesting meadows in this way, to keep off all sorts of pests."

Organisms of the fungus type include the so-called *fungi imperfecti* and a botanical transition-form, the family of actinomycetes and streptomycetes, from which certain antibiotic drugs are derived. I have found that these organisms play a very special rôle in humus formation and decay, and that they are abundantly present in the bio-dynamic manure and compost preparations. The preparations also contain an abundance of many of the most important trace elements, such as molybdenum, cobalt, zinc, and others whose importance has been experimentally demonstrated.

Now a peculiar situation was found to exist in regard to soils. Analyses of available plant nutrients showed that the same soil tested quite differently at different seasons. Indeed, tests showed not only seasonal but even daily variations. The same soil sample often disclosed periodic variations greater than those found in tests of soils from adjoining fields, one of which was good, the other poor. Seasonal and daily variations are influenced, however, by the earth's relative position in the planetary system; they are, in other words, of cosmic origin. It has actually been found that the time of day or the season of the year influences the solubility and availability of nutrient substances. Numerous phenomena to be observed in the physiology of plants and animals (*e.g.* glandular secretions, hormones) are subject to such influences. The concentration of oxalic acid in bryophyllum leaves rises and falls with the time of day with almost clock-like regularity. Although in this and many other test cases the nutrients on which the plants were fed were identical, the increase or decrease in the plant's substantial content varied very markedly in response to varying light-rhythms and cycles. Joachim Schultz, a research worker at the Goetheanum whose life was most unfortunately cut short, had begun to test Dr. Steiner's important indication that light activity acts with growth-stimulating effect in the morning and late afternoon hours, while at noon and midnight its influence is growth-inhibiting.

When I inspected Schultz's experiments, I was struck by the fact that plants grown on the same nutrient solution had a wholly different substantial composition according to the light-rhythms operative. This was true of nitrogen, for example. Plants exposed to light during the morning and evening hours grew strongly under the favourable influence of nitrogen activity, whereas if exposed during the noon hours, they declined and showed deficiency symptoms. The way was thus opened for experimental demonstration of the fact that the so-called "cosmic" activity of light, of warmth, of sun forces especially, but of other light-sources also, prevails over the material processes. These cosmic forces regulate the course of material change. When and in what direction this takes place, and the extent to which the total growth and the form of the plant are influenced, all depend upon the cosmic constellation and the origin of the forces concerned. Recent research in the field of photosynthesis has produced findings which can hardly fail to open the eyes even of materialistic observers to such processes. Here, too, Rudolf Steiner is shown to have been a pioneer who paved the way for a new direction of research. It is impossible in an article of this length to report on all the phenomena that have already been noted, for they would more than fill a book. But it is no longer possible to dismiss the influence of cosmic forces as "mere superstition" when the physiological and biochemical inter-relationships of metabolic functions in soil-life, the rise and fall of sap in the plant, and especially processes in the root-sphere are taken into consideration.

* * *

In an earlier view of nature, based partly on old mystery-tradition and partly on instinctive clairvoyance—a view originating in the times of Aristotle and his pupil Theophrastus, and continuing on to the days of Albertus Magnus and the late mediaeval "doctrine of signatures"—it was recognised that relationships exist between certain cosmic constellations and the various plant species. These constellations are creative moments under whose influence species became differentiated and the various plant forms came into being. When one realises that cosmic rhythms have such a significant influence on the physiology of metabolism, of glandular functions, of the rise and fall of sap and of sap pressure (turgor), only a small step remains to be taken by conscious future research to the next realisation, which will achieve an experimental grasp of these creative constellations. Many of Rudolf Steiner's collaborators have already demonstrated the decisive effects of formative forces in such experiments as the capillary tests on filter paper of L. Kolisko and the plant and crystallisation tests of Pfeiffer, Krüger, Bessenich, Selawry and others.

Rudolf Steiner's suggestions for plant breeding presented a special task. Research in this field was carried out by the author and other fellow-workers (Immanuel Vögele, Erika Riese, Martha Kuenzel and Martin Schmidt), either in collaboration or in independent work. Proceeding from the basic concept of creative cosmic constellations, one can assume that the original creative impetus in every species of sub-type slowly exhausts itself and ebbs away. The formative forces of this original impulse is passed on from plant to plant in hereditary descent by means of certain organs such as chromosomes. One-sided quantity-manuring gradually inhibits the activity of the primary forces, and results in a weakening of the plant. Seed quality degenerates. This was the initial problem laid before Rudolf Steiner, and the bio-dynamic movement came into being as an answer to it.

The task was to reunite the plant, viewed as a system of forces under the influence of cosmic activities, with nature as a whole. Rudolf Steiner pointed out that many plants which had been "violated," in the sense of having been estranged from their cosmic origin, were already so far gone in degeneration that by the end of the century their propagation would be unreliable. Wheat and potatoes were among the plant types mentioned, but other such grains as oats, barley and lucerne belong to the same picture. Ways were sketched whereby new strains with strong seed-forces could be bred from "unexhausted" relatives of the cultivated plants. This work has begun to have success; the species of wheat have already been developed. Martin Schmidt carried on significant researches, not yet published, to determine the rhythm of seed placement in the ear, and to show in particular the difference between food plants and plants grown for seed. According to Rudolf Steiner, there is a basic difference between the two types, one of which is sown in autumn, nearer to the winter, and the other nearer to the summer. Biochemists will eventually be able to confirm these differences materially in the structure of protein substances, amino-acids, phosphorlipoids, enzyme-systems and so on by means of modern chromatographic methods.

The degeneration of wheat is already an established fact. Even where the soil is good, the protein content has declined; in the case of soft red wheat, protein content has sunk from 13% to 8% in some parts of the United States. Potato growers know how hard it is to produce healthy potatoes free from viruses and insects, not to mention the matter of flavour. Bio-dynamically grown wheat maintains its high protein level. Promising work in potato breeding was unfortunately interrupted by the last war and other disturbances.

Pests are one of the most interesting and instructive problems, looked at from the bio-dynamic viewpoint. When the biological balance is upset, degeneration follows; pests and diseases make their appearance. Nature herself liquidates weaklings. Pests are therefore to be regarded as nature's warning that the primary forces have been dissipated and the balance sinned against. According to official estimates, American agriculture pays a yearly bill of five thousand million dollars in crop losses for disregarding this warning, and another seven hundred and fifty million dollars on keeping down insect pests. People are beginning to realise that insect poisons fall short of solving the problem, especially since the destruction of some of the insects succeeds only in producing new, more resistant kinds. It has been established by the most advanced research (Albrecht of Missouri) that one-sided fertilising disturbs the protein-carbohydrates balance in plant cells, to the detriment of proteins and the layer of wax that coats plant leaves, and makes the plants "tastier" to insect depredators. It has been a bitter realisation that insect poisons merely "preserve" a part of moribund nature, but do not halt the general trend towards death. Experienced entomologists, who have witnessed the failure of chemical pest-control and the threats to health associated with it, are beginning to speak out and demand biological controls. But according to the findings of one of the American experimental stations, biological controls are feasible only when no poisons are used and an attempt is made to restore natural balance. In indications given in the Agriculture Course, Rudolf Steiner showed that health and resistance are functions of biological balance, coupled with cosmic factors. This is further evidence of how far in advance of its time was this spiritual-scientific, Goethean way of thought.

The author is thoroughly conscious of the fact that this exposition touches upon only a small part of the whole range of questions opened up by Rudolf Steiner's new agricultural method. He is also aware that other collaborators would have written quite differently, and about different aspects of the work. These pages should therefore be read in accordance with their intention: as the view from a single window in a house containing many rooms.

(Contributed by Dr. Pfeiffer to the German symposium, *Wir erlebten Rudolf Steiner*, of which a complete English translation, "Rudolf Steiner, by his pupils," was published as a special number of *The Golden Blade*, 1958. This translation is used by permission of *The Golden Blade* and the Verlag Freies Geistesleben G.m.b.H., Stuttgart, publishers of the book, *Wir erlebten Rudolf Steiner*.)

LECTURE ONE

KOBERWITZ,
7th June, 1924.

MY DEAR FRIENDS,

With profound thanks I look back on the words which Count Keyserlingk has just spoken. For the feeling of thanks is not only justified on the part of those who are able to receive from Anthroposophical Science. One can also feel deeply what I may call the thanks of Anthroposophia itself—thanks which in these hard times are due to all who share in anthrosposophical interests.

Out of the spirit of Anthroposophia, therefore, I would thank you most heartily for the words you have just spoken. Indeed, it is deeply gratifying that we are able to hold this Agriculture Course here in the house of Count and Countess Keyserlingk. I know from my former visits what a beautiful atmosphere there is in Koberwitz—I mean also the spiritual atmosphere. I know that the atmosphere of soul and spirit which is living here is the best possible premiss for what must be said during this Course.

Count Keyserlingk has told us that there may be some discomforts for one or another among us. He was speaking especially of the eurhythmists; though it may be the "discomforts" are shared by some of our other visitors from a distance. Yet on the other hand, considering the purpose of our present gathering, it seems to me we could scarcely be accommodated better for this Lecture Course than here, in a farm so excellent and so exemplary.

Whatever comes to light in the realms of Anthroposophia, we also need to live in it with our feelings—in the necessary atmosphere. And for our Course on Farming this condition will most certainly be fulfilled at Koberwitz. All this impels me to express our deeply felt thanks to Count Keyserlingk and to his house. In this I am sure Frau Doctor Steiner will join me. We are thankful that we may spend these festive days—I trust they will also be days of real good work—here in this house.

I cannot but believe: inasmuch as we are gathered here in Koberwitz, there will prevail throughout these days an agricultural spirit which is already deeply united with the Anthroposophical Movement. Was it not Count Keyserlingk who helped us from the very outset with his advice and his devoted work, in the farming activities we undertook at Stuttgart under the Kommende Tag Company? His spirit, trained by his deep and intimate union with Agriculture, was prevalent in all that we were able to do in this direction. And I would say, forces were there prevailing which came from the innermost heart of our Movement and which drew us hither, quite as a matter of course, the moment the Count desired us to come to Koberwitz.

Hence I can well believe that every single one of us has come here gladly for this Agriculture Course. We who have come here can express our thanks just as deeply and sincerely, that your House has been ready to receive us with our intentions for these days. For my part, these thanks are felt most deeply, and I beg Count Keyserlingk and his whole house to receive them especially from me. I know what it means to give hospitality to so many visitors and for so many days, in the way in which I feel it will be done here. Therefore I think I can also give the right colouring to these words of thanks, and I beg you to receive them, understanding that I am well aware of the many difficulties which such a gathering may involve in a house remote from the city. Whatever may be the inconveniences of which the Count has spoken—representing, needless to say, not the "Home Office" but the "Foreign Office"—whatever they may be, I am quite sure that every single one of us will go away fully satisfied with your kind hospitality.

Whether you will go away equally satisfied with the Lecture-Course itself, is doubtless a more open question, though we will do our utmost, in the discussions during the succeeding days, to come to a right understanding on all that is here said. You must not forget: though the desire for it has been cherished in many quarters for a long time past, this is the first time I have been able to undertake such a Course out of the heart of our anthroposophical striving. It pre-supposes many things.

The Course itself will show us how intimately the interests of Agriculture are bound up, in all directions, with the widest spheres of life. Indeed there is scarcely a realm of human life which lies outside our subject. From one aspect or another, all interests of human life belong to Agriculture. Here, needless to say, we can only touch upon the central domain of Agriculture itself, albeit this of its own accord will lead us along many different side tracks—necessarily so, for the very reason that what is here said will grow out of the soil of Anthroposophia itself.

In particular, you must forgive me if my introductory words to-day appear—inevitably—a little far remote. Not everyone, perhaps, will see at once what the connection is between this introduction and our special subject. Nevertheless, we shall have to build upon what is said to-day, however remote it may seem at first sight. For Agriculture especially is sadly hit by the whole trend of modern spiritual life. You see, this modern spiritual life has taken on a very destructive form especially as regards the economic realm, though its destructiveness is scarcely yet divined by many.

Our real underlying intentions, in the economic undertakings which grew out of the Anthroposophical Movement, were meant to counteract these things. These undertakings were created by industrialists, business men, but they were unable to realise in all directions what lay in their original intentions, if only for the reason that the opposing forces in our time are all too numerous, preventing one from calling forth a proper understanding for such efforts. Over against the "powers that be," the individual is often powerless. Hitherto, not even the most original and fundamental aspects of these industrial and economic efforts, which grew out of the heart of the Anthroposophical Movement, have been realised. Nay, they have not even reached the plane of discussion. What was the real, practical point? I will explain it in the case of Agriculture, so that we may not be speaking in vague and general, but in concrete terms.

We have all manner of books and lecture courses on Economics, containing, among other things, chapters on the economic aspects of Agriculture. Economists consider, how Agriculture should be carried on in the light of social-economic principles. There are many books and pamphlets on this subject: how Agriculture should be shaped, in the light of social and economic ideas. Yet the whole of this—the giving of economic lectures on the subject and the writing of such books—is manifest nonsense. Palpable nonsense, I say, albeit that is practised nowadays in the widest circles. For it *should* go without saying, and every man should recognise the fact : One cannot speak of Agriculture, not even of the social forms it should assume, unless one first possesses as a foundation a practical acquaintance with the farming job itself. That is to say, unless one really knows what it means to grow mangolds, potatoes and corn! Without this foundation one cannot even speak of the general economic principles which are involved. Such things must be determined out of the thing itself, not by all manner of theoretic considerations.

Nowadays, such a statement seems absurd to those who have heard University lectures on the economics of Agriculture. The whole thing seems to them so well established. But it is not so. No one can judge of Agriculture who does not derive his judgment from field and forest and the breeding of cattle. All talk of Economics which is not derived from the job itself should really cease. So long as people do not recognise that all talk of Economics—hovering airily over the realities—is mere empty talk, we shall not reach a hopeful prospect, neither in Agriculture nor in any other sphere.

Why is it that people think they can talk of a thing from theoretic points of view, when they do not understand it? The reason is, that even within their several domains they are no longer able to go back to the real foundations. They look at a beetroot as a beetroot. No doubt it has this or that appearance; it can be cut more or less easily, it has such and such a colour, such and such constituents. All these things can no doubt be said. Yet therewithal you are still far from understanding the beetroot. Above all, you do not yet understand the living-together of the beetroot with the soil, with the field, the season of the year in which it ripens, and so forth.

You must be clear as to the following (I have often used this comparison for other spheres of life): You see a magnetic needle. You discern that it always points with one end approximately to the North, and with the other to the South. You think, why is it so? You look for the cause, not in the magnetic needle, but in the whole Earth, inasmuch as you assign to the one end of the Earth the magnetic North Pole, and to the other the magnetic South.

Anyone who looked in the magnet-needle itself for the cause of the peculiar position it takes up, would be talking nonsense. You can only understand the direction of the magnet-needle if you know how it is related to the whole Earth. Yet the same nonsense (as applied to the magnetic needle) is considered good sense by the men of to-day when applied to other things.

There, for example, is the beetroot growing in the earth. To take it just for what it is within its narrow limits, is nonsense if in reality its growth depends on countless conditions, not even only of the Earth as a whole, but of the cosmic environment. The men of to-day say and do many things in life and practice as though they were dealing only with narrow, limited objects, not with effects and influences from the whole Universe. The several spheres of modern life have suffered terribly from this, and the effects would be even more evident were it not for the fact that in spite of all the modern science a certain instinct still remains over from the times when men were used to work by instinct and not by scientific theory.

To take another sphere of life: I am always glad to think that those whose doctors have prescribed how many ounces of meat they are to eat, and how much cabbage (some of them even have a balance beside them at the table and carefully weigh out everything that comes on to their plate)—it is all very nice; needless to say, one ought to know such things—but I am always glad to think how good it is that the poor fellow still feels hungry, if, after all, he has not had enough to eat! At least there is still this instinct to tell him so.

Such instincts really underlay all that men had to do before a "science" of these things existed. And the instincts frequently worked with great certainty. Even to-day one is astonished again and again to read the rules in the old "Peasants' Calendars." How infinitely wise and intelligent is that which they express! Moreover, the man of sure instincts is well able to avoid superstition in these matters: and in these Calendars, beside the proverbs full of deep meaning for the sowing and the reaping, we find all manner of quips, intended to set aside nonsensical pretentions. This for example:—

"Kräht der Hahn auf dem Mist,
So regnet es, oder es bleibt wie es ist."

"If the cock crows on the dunghill,
It'll rain—or it'll stay still."

So the needful dose of humour is mingled with the instinctive wisdom in order to ward off mere superstition.

We, however, speaking from the point of view of Anthroposophical Science, do not desire to return to the old instincts. We want to find, out of a deeper spiritual insight, what the old instincts—as they are growing insecure—are less and less able to provide. To this end we must include a far wider horizon in our studies of the life of plant and animal, and of the Earth itself. We must extend our view to the whole Cosmos.

From one aspect, no doubt, it is quite right that we should not superficially connect the rain with the phases of the Moon. Yet on the other hand there is a true foundation to the story I have often told in other circles. In Leipzig there were two professors. One of them, Gustav Theodor Fechner, often evinced a keen and sure insight into spiritual matters. Not altogether superstitiously, from pure external observations he could see that certain periods of rain or of no rain were connected, after all, with the Moon and with its coursing round the earth.

He drew this as a necessary conclusion from the statistical results. That however was a time when orthodox science already wanted to overlook such matters, and his colleague, the famous Professor Schleiden, poured scorn on the idea "for scientific reasons." Now these two professors of the University of Leipzig also had wives. Gustav Theodor Fechner, who was a man not without humour, said: "Well, let our wives decide."

In Leipzig at that time the water they needed for washing clothes was not easy to obtain, and a certain custom still prevailed. You had to fetch your water from a long distance. Hence they were wont to put out pails and barrels to catch the rain water.

This was Frau Prof. Schleiden's custom as well as Frau Prof. Fechner's. But they had not room enough to put out their barrels in the yard at the same time. So Prof. Fechner said: "If my honoured colleague is right, if it makes no difference, then let Frau Prof. Schleiden put out her barrel when by my indications, according to the phases of the Moon, there will be less rain. If it is all nonsense, Frau Prof. Schleiden will surely be glad to do so."

But, lo and behold, Frau Prof. Schleiden rebelled. She preferred the indications of Prof. Fechner to those of her own husband. And so indeed it is. Science may be perfectly correct. Real life, however, often cannot afford to take its cue from the "correctness" of science!

But we do not wish to speak only in this way. We are in real earnest about it. I only wanted to point out the need to look a little farther afield than is customary nowadays. We *must* do so in studying that which alone makes possible the physical life of man on Earth—and that, after all, is Agriculture. I do not know whether the things which can be said at this stage out of Anthroposophical Science will satisfy you in all directions, but I will do my best to explain what Anthroposophical Science can give for Agriculture.

* * *

To-day, by way of introduction, I will indicate what is most important for Agriculture in the life of the Earth. Nowadays we are wont to attach the greatest importance to the physical and chemical constituents. To-day, however, we will not take our start from these; we will take our start from something which lies *behind* the physical and chemical constituents and is nevertheless of great importance for the life of plant and animal.

Studying the life of man (and to a certain extent it applies to animal life also), we observe a high degree of emancipation of human and animal life from the outer Universe. The nearer we come to man, the greater this emancipation grows. In human and animal life we find phenomena appearing—to begin with—quite independent not only of the influences from beyond the Earth, but also of the atmospheric and other influences of the Earth's immediate environment. Moreover, this not only appears so; it is to a high degree correct for many things in human life.

True, it is well-known that the pains of certain illnesses are intensified by atmospheric influences. There is, however, another fact of which the people of to-day are not so well aware. Certain illnesses and other phenomena of human life take their course in such a way that in their time-relationships they copy the external processes of Nature. Yet in their beginning and end they do not coincide with these Nature-processes. We need only call to mind one of the most important phenomena of all, that of female menstruation. The periods, in their temporal course, imitate the course of the lunar phases, but they do not coincide with the latter in their beginning and ending. And there are many other, less evident phenomena, both in the male and in the female organism, representing imitations of rhythms in outer Nature.

If these things were studied more intimately, we should for example have a better understanding of many things that happen in the social life by observing the periodicity of the Sun-spots. People only fail to observe these things because that in human life which corresponds to the periodicity of the Sun-spots does not begin when they begin, nor does it cease when they cease. It has emancipated itself. It shows the same periodicity, the identical rhythm, but its phases do not coincide in time. While inwardly maintaining the rhythm and periodicity, it makes them independent—it emancipates itself.

Anyone, of course, to whom we say that human life is a microcosm and imitates the macrocosm, is at liberty to reply. That is all nonsense! If we declare that certain illnesses show a seven day's fever period, one may object: Why then, when certain outer phenomena appear, does not the fever too make its appearance and run parallel, and cease with the external phenomena? It is true that the fever does not; but, though its temporal beginning and ending do not coincide with the outer phenomena, it still maintains their inner rhythm. This emancipation in the Cosmos is almost complete for human life; for animal life it is less so; plant life, on the other hand, is still to a high degree immersed in the general life of Nature, including the outer earthly world.

Hence we shall never understand plant life unless we bear in mind that everything which happens on the Earth is but a reflection of what is taking place in the Cosmos. For man this fact is only masked because he has emancipated himself; he only bears the *inner* rhythms in himself. To the plant world, however, it applies in the highest degree. That is what I should like to point out in this introductory lecture.

The Earth is surrounded in the heavenly spaces, first by the Moon and then by the other planets of our planetary system. In an old instinctive science wherein the Sun was reckoned among the planets, they had this sequence: Moon, Mercury, Venus, Sun, Mars, Jupiter, Saturn. Without astronomical explanations I will now speak of this planetary life, and of that in the planetary life which is connected with the earthly world.

Turning our attention to the earthly life on a large scale, the first fact for us to take into account is this. The greatest imaginable part is played in this earthly life (considered once more on a large scale, and as a whole) by all that which we may call the life of the *silicious substance* in the world. You will find silicious substance for example, in the beautiful mineral quartz, enclosed in the form of a prism and pyramid; you will find the silicious substance, combined with oxygen, in the crystals of quartz.

Imagine the oxygen removed (which in the quartz is combined with silicious substance) and you have so-called silicon. This substance is included by modern chemistry among the "elements," oxygen, nitrogen, hydrogen, sulphur, etc. Silicon therefore, which is here combined with oxygen, is a "chemical element."

Now we must not forget that the silicon which lives thus in the mineral quartz is spread over the Earth so as to constitute 27-28% of our Earth's crust. All other substances are present in lesser quantities, save oxygen, which constitutes 47-48%. Thus an enormous quantity of silicon is present. Now, it is true this silicon, occurring as it does in rocks like quartz, appears in such a form that it does not seem very important when we are considering the outer, material aspect of the Earth with its plant-growth. (The plant-growth is frequently forgotten).

Quartz is insoluble in water—the water trickles through it. It therefore seems—at first sight—to have very little to do with the ordinary, obvious conditions of life. But once again, you need only remember the horse-tail—equisetum—which contains 90% of silica—the same substance that is in quartz—very finely distributed.

From all this you can see what an immense significance silicon must have. Well-nigh half of what we meet on the Earth consists of silica. But the peculiar thing is how very little notice is taken of it. It is practically excluded to-day even from those domains of life where it could work most beneficially.

In the Medicine that proceeds from Anthroposophical Science, silicious substances are an essential constituent of numerous medicaments. A large class of illnesses are treated with silicic acid taken internally, or outwardly as baths. In effect, practically everything that shows itself in abnormal conditions of the senses is influenced in a peculiar way by silicon. (I do not say what lies in the senses themselves, but that which *shows itself* in the senses, including the inner senses—calling forth pains here or there in the organs of the body).

Not only so; throughout the "household of Nature," as we have grown accustomed to call it, silicon plays the greatest imaginable part, for it not only exists where we discover it in quartz or other rocks, but in an extremely fine state of distribution it is present in the atmosphere. Indeed, it is everywhere. Half of the Earth that is at our disposal is of silica.

Now what does this silicon do? In a hypothetical form, let us ask ourselves this question. Let us assume that we only had half as much silicon in our earthly environment. In that case our plants would all have more or less pyramidal forms. The flowers would all be stunted. Practically all plants would have the form of the cactus, which strikes us as abnormal. The cereals would look very queer indeed. Their stems would grow thick, even fleshy, as you went downward; the ears would be quite stunted—they would have no full ears at all.

That on the one hand. On the other hand we find another kind of substance, which must occur everywhere throughout the Earth, albeit it is not so widespread as the silicious element. I mean the chalk or limestone substances and all that is akin to these—limestone, potash, sodium substances. Once more, if these were present to a less extent, we should have plants with very thin stems—plants, to a large extent, with twining stems; they would all become like creepers. The flowers would expand, it is true, but they would be useless: they would provide practically no nourishment. Plant-life in the form in which we see it to-day can only thrive in the equilibrium and co-operation of the two forces—or, to choose two typical substances, in the co-operation of the limestone and silicious substances respectively.

Now we can go still farther. Everything that lives in the silicious nature contains forces which comes not from the Earth but from the so-called *distant planets*, the planets beyond the Sun—Mars, Jupiter and Saturn. That which proceeds from these distant planets influences the life of plants via the silicious and kindred substances into the plant and also into the animal life of the Earth. On the other hand, from all that is represented by the planets *near* the Earth—Moon, Mercury and Venus—forces work via the limestone and kindred substances. Thus we may say, for every tilled field: Therein are working the silicious and the limestone natures; in the former, Saturn, Jupiter and Mars; and in the latter, Moon, Venus and Mercury.

In this connection let us now look at the plants themselves. Two things we must observe in the plant life. The first thing is that the entire plant-world, and every single species, is able to maintain iself —that is to say, it evolves the power of reproduction. The plant is able to bring forth its kind, and so on. That is the one thing. The other is, that as a creature of a comparatively lower kingdom of Nature, the plant can serve as nourishment for those of the higher kingdoms.

At first sight, these two currents in the life and evolution of the plant have little to do with one another. For the process of development from the mother plant to the daughter plant, the grand-daughter plant and so on, it may well seem a matter of complete indifference to the formative forces of Nature, whether or no we eat the plant and nourish ourselves thereby. Two very different sets of interests are manifested here. Yet in the whole nexus of Nature's forces, it works in this way:—

Everything connected with the inner force of reproduction and growth—everything that contributes to the sequence of generation after generation in the plants—works through those forces which come down from the Cosmos to the Earth: from Moon, Venus and Mercury, via the limestone nature. Suppose we were merely considering what emerges in plants such as we do not eat—plants that simply renew themselves again and again. We look at them as though the cosmic influences from the forces of Venus, Mercury and Moon did not interest us. For these are the forces involved in all that reproduces itself in the plant-nature of the Earth.

On the other hand, when plants become foodstuffs to a large extent—when they evolve in such a way that the substances in them become foodstuffs for animal and man, then Mars, Jupiter and Saturn, working via the silicious nature, are concerned in the process. The silicious nature opens the plant-being to the wide spaces of the Universe and awakens the senses of the plant-being in such a way as to receive from all quarters of the Universe the forces which are moulded by these distant planets. Whenever this occurs, Mars, Jupiter and Saturn are playing their part. From the sphere of the Moon, Venus and Mercury, on the other hand, is received all that which makes the plant capable of reproduction.

To begin with, no doubt this appears as a simple piece of information. But truths like this, derived from a somewhat wider horizon, lead of their own accord from knowledge into practice. For we must ask ourselves: If forces come into the Earth from Moon, Venus and Mercury and become effective in the life of plants, by what means can the process be more or less quickened or restrained? By what means can the influences of Moon or Saturn on the life of plants be hindered, and by what means assisted?

Observe the course of the year. It takes its course in such a way that there are days of rain and days without rain. As to the rain, the modern physicist investigates practically no more than the mere fact that when it rains, more water falls upon the Earth than when it does not rain. For him, the water is an abstract substance composed of hydrogen and oxygen. True, if you decompose water by electrolysis, it will fall into two substances, of which the one behaves in such and such a way, and the other in another way. But that does not yet tell us anything complete about water itself. Water contains far, far more than what emerges from it chemically, in this process, as oxygen and hydrogen.

Water, in effect, is eminently suited to prepare the ways within the earthly domain for those forces which come, for instance, from the Moon. Water brings about the distribution of the lunar forces in the earthly realm. There is a definite connection between the Moon and the water in the Earth. Let us therefore assume that there have just been rainy days and that these are followed by a full Moon. In deed and in truth, with the forces that come from the Moon on days of the full Moon, something colossal is taking place on Earth. These forces spring up and shoot into all the growth of plants, but they are unable to do so unless rainy days have gone before.

We shall therefore have to consider the question: Is it not of some significance, whether we sow the seed in a certain relation to the rainfall and the subsequent light of the full Moon, or whether we sow it thoughtlessly at any time? Something, no doubt, will come of it even then. Nevertheless, we have to raise this question: How should we best consider the rainfall and the full Moon in choosing the time to sow the seed? For in certain plants, what the full Moon has to do will thrive intensely after rainy days and will take place but feebly and sparingly after days of sunshine. Such things lay hidden in the old farmers' rules; they quoted a certain verse or proverb and knew what they must do. The proverbs to-day are outworn superstitions, and a science of these things does not yet exist; people are not yet willing enough to set to work and find it.

Furthermore, around our Earth is the atmosphere. Now the atmosphere above all—beside the obvious fact that it is airy—has the peculiarity that it is sometimes warmer, sometimes cooler. At certain times it shows a considerable accumulation of warmth, which, when the tension grows too strong, may even find relief in thunderstorms. How is it then with the *warmth*? Spiritual observation shows that whereas the water has no relation to *silica*, this warmth has an exceedingly strong relation to it.

The warmth brings out and makes effective precisely those forces which can work through the silicious nature, namely, the forces that proceed from Saturn, Jupiter and Mars. These forces must be regarded in quite a different way than the forces from the Moon. For we must not forget that Saturn takes thirty years to revolve round the Sun, whereas the Moon with its phases takes only thirty or twenty-eight days. Saturn is only visible for fifteen years. It must therefore be connected with the growth of plants in quite a different way, albeit, I need hardly say, it is not only working when it shines down upon the Earth; it is also effective when its rays have to pass upward through the Earth.

Saturn goes slowly round, in thirty years. Let us draw it thus (Diagram 1): here is the course of Saturn. Sometimes it shines directly on to a given spot of the Earth. But it can also work *through* the Earth upon this portion of the Earth's surface. In either case the intensity with which the Saturn-forces are able to approach the plant life of the Earth is dependent on the warmth-conditions of the air. When the air is cold, they cannot approach; when the air is warm, they can.

And where do we see the working of these forces in the plant's life? We see it, not so much where annual plants arise, coming and going in a season and only leaving seeds behind. We see what Saturn does with the help of the warmth-forces of our Earth, whenever the perennial plants arise. The effects of these forces, which pass into the plant-nature via the warmth, are visible to us in the rind and bark of trees, and in all that makes the plants, perennial. This is due to the simple fact that the annual life of the plant—its limitation to a short length of life—is connected with those planets whose period of revolution is short. That, on the other hand, which frees itself from the transitory nature—that which surrounds the trees with bark and rind, and makes them permanent—is connected with the planetary forces which work via the forces of warmth and cold and have a long period of revolution, as in the case of Saturn: thirty years; or Jupiter: twelve years.

If someone wishes to plant an oak, it is of no little importance whether or no he has a good knowledge of the periods of Mars; for an oak, rightly planted in the proper Mars-period, will thrive differently from one that is planted in the Earth thoughtlessly, just when it happens to suit.

Or, if you wish to plant coniferous forests, where the Saturn-forces play so great a part, the result will be different if you plant the forest in a so-called ascending period of Saturn, or in some other Saturn period. One who understands can tell precisely, from the things that will grow or will not grow, whether or no they have been planted with an understanding of the connections of these forces. That which does not appear obvious to the external eye, appears very clearly, none the less, in the more intimate relationships of life.

Assume for instance that we take, as firewood, wood that is derived from trees which were planted in the Earth without understanding of the cosmic rhythms. It will not provide the same health-giving warmth as firewood from trees that were planted intelligently. These things enter especially into the more intimate relationships of daily life, and here they show their great significance. Alas! the life of people has become almost entirely thoughtless nowadays. They are only too glad if they do not need to think of such things. They think it must all go on just like any machine. You have all the necessary contrivances; turn on the switch, and it goes. So do they conceive, materialistically, the working of all Nature.

Along these lines we are eventually led to the most alarming results in practical life. Then the great riddles arise. Why, for example, is it impossible to-day to eat such potatoes as I ate in my youth? It is so; I have tried it everywhere. Not even in the country districts where I ate them then, can one now eat such potatoes. Many things have declined in their inherent food-values, notably during the last decades.

The more intimate influences which are at work in the whole Universe are no longer understood. These must be looked for again along such lines as I have hinted at to-day. I have only introduced the subject; I have only tried to show where the questions arise—questions which go far beyond the customary points of view. We shall continue and go deeper in this way, and then apply, what we have found, in practice.

LECTURE TWO

KOBERWITZ,
10th June, 1924.

MY DEAR FRIENDS,

We shall spend the first lectures gathering various items of knowledge, so as to recognise the conditions on which the prosperity of Agriculture depends. Thereafter we shall draw the practical conclusions, which can only be realised in the immediate application and are only significant when put into practice. In these first lectures you must observe how all agricultural products arise; how Agriculture lives in the totality of the Universe.

A farm is true to its essential nature, in the best sense of the word, if it is conceived as a kind of individual entity in itself—a self-contained individuality. Every farm should approximate to this condition. This ideal cannot be absolutely attained, but it should be observed as far as possible. Whatever you need for agricultural production, you should try to posses it within the farm itself (including in the "farm," needless to say, the due amount of cattle). Properly speaking, any manures or the like which you bring into the farm from outside should be regarded rather as a remedy for a sick farm. That is the ideal. A thoroughly healthy farm should be able to produce within itself all that it needs.

We shall see presently why this is the natural thing. So long as one does not regard things in their true essence but only in their outer material aspect, the question may justifiably arise: Is it not a matter of indifference whether we get our cow-dung from the neighbourhood or from our own farm? But it is not so. Although these things may not be able to be strictly carried out, nevertheless, if we wish to do things in a proper and natural way, we need to have this ideal concept of the necessary self-containedness of any farm.

You will recognise the justice of this statement if you consider the Earth on the one hand, from which our farm springs forth, and on the other hand, that which works down into our Earth from the Universe beyond. Nowadays, people are wont to speak very abstractly of the influences which work on to the Earth from the surrounding Universe. They are aware, no doubt, that the Sun's light and warmth, and all the meteorological processes connected with it, are in a way related to the form and development of the vegetation that covers the soil. But present-day ideas can give no real information as to the exact relationships, because they do not penetrate to the realities involved. We shall have to consider the matter from various standpoints. Let us to-day choose this one: let us consider, to begin with, the soil of the Earth which is the foundation of all Agriculture.

I will indicate the surface of the Earth diagramatically by this line (Diagram 2). The surface of the Earth is generally regarded as mere mineral matter—including some organic elements, at most, inasmuch as there is formation of humus, or manure is added. In reality, however, the earthly soil *as such* not only contains a certain *life*—a vegetative nature of its own—but an effective *astral principle* as well; a fact which is not only not taken into account to-day but is not even admitted nowadays. But we can go still further. We must observe that this inner life of the earthly soil (I am speaking of fine and intimate effects) is different in summer and in winter. Here we are coming to a realm of knowledge, immensely significant for practical life, which is not even thought of in our time.

Taking our start from a study of the earthly soil, we must indeed observe that the surface of the Earth is a kind of organ in that organism which reveals itself throughout the growth of Nature. The Earth's surface is a real organ, which—if you will—you may compare to the human diaphragm. (Though it is not quite exact, it will suffice us for purposes of illustration). We gain a right idea of these facts if we say to ourselves: Above the human diaphragm there are certain organs—notably the head and the processes of breathing and circulation which work up into the head. Beneath it there are other organs.

If from this point of view we now compare the Earth's surface with the human diaphragm, then we must say: In the individuality with which we are here concerned, the head is *beneath* the surface of the Earth, while we, with all the animals, are living in the creature's belly! Whatever is *above* the Earth, belongs in truth to the intestines of the "agricultural individuality," if we may coin the phrase. We, in our farm, are going about in the belly of the farm, and the plants themselves grow upward in the belly of the farm. Indeed, we have to do with an individuality standing on its head. We only regard it rightly if we imagine it, compared to man, as standing on its head. With respect to the animal, as we shall presently see, it is a little different.

Why do I say that the agricultural individuality is standing on its head? For the following reason. Take everything there is in the immediate neighbourhood of the Earth by way of air and water-vapours and even warmth. Consider, once more, all that element in the neighbourhood of the Earth in which we ourselves are living and breathing and from which the plants, along with us, receive their outer warmth and air, and even water. All this actually corresponds to that which would represent, in man, the abdominal organs. On the other hand, that which takes place in the interior of the Earth—beneath the Earth's surface—works upon plant-growth in the same way in which our head works upon the rest of our organism, notably in childhood, but also throughout our life. There is a constant and living mutual interplay of the above-the-Earth and the below-the-Earth.

And now, to localise these influences, I beg you to observe the following. The activities above the Earth are immediately dependent on Moon, Mercury and Venus supplementing and modifying the influences of the Sun. The so-called "planets near the Earth" extend their influences to all that is *above* the Earth's surface. On the other hand, the distant planets—those that revolve outside the circuit of the Sun—work upon all that is *beneath* the Earth's surface, assisting those influences which the Sun exercises from below the Earth. Thus, so far as plant-growth is concerned, we must look for the influences of the distant Heavens *beneath*, and of the Earth's immediate cosmic environment *above* the Earth's surface.

Once more: all that works inward from the far spaces of the Cosmos to influence the growth of plants, works not directly—not by direct radiation—but in this way: It is first received by the Earth, and the Earth then rays it upward again. Thus, the influences that rise upward from the earthly soil—beneficial or harmful for the growth of plants—are in reality cosmic influences rayed back again and working directly in the air and water over the Earth. The direct radiation from the Cosmos is stored up beneath the Earth's surface and works back from thence. Now these relationships determine how the earthly soil, according to its constitution, works upon the growth of plants. (We shall take plant-growth to begin with, and afterwards extend it to the animals).

Consider the earthly soil. To begin with, we have those influences that depend on the farthest distances of the Cosmos—the farthest that come into account for earthly processes. These effects are found in what is commonly called sand and rock and stone. Sand and rock—substances impermeable to water, which, in the common phrase, "contain no foodstuffs"—are in reality no less important than any other factors. They are most important for the unfolding of the growth-processes, and they depend throughout on the influences of the most distant cosmic forces. And above all—improbable as it appears at first sight—it is through the sand, with its *silicious* content, that there comes into the Earth what we may call the *life-ethereal* and the *chemically* influential elements of the soil. These influences then take effect as they ray upward again from the Earth.

The way the soil itself grows inwardly alive and develops its own chemical processes, depends above all on the composition of the sandy portion of the soil. What the plant-roots experience in the soil depends in no small measure on the extent to which the cosmic life and cosmic chemistry are seized and held by means of the stones and the rock, which may well be at a considerable depth beneath the surface. Therefore, wherever we are studying plant-growth, we should be clear in the first place as to the geological foundation out of which it arises. For those plants in which the root-nature as such is important, we should never forget that a silicious ground—even if it be only present in the depths below—is indispensable. I would say, thanks be to God that silica is very widespread on the Earth—in the form of silicic acid, for instance, and in other compounds. It constitutes 47-48% of the surface of the Earth, and for the quantities we need we can reckon practically everywhere on the presence of the silicic activity.

But that is not all. All that is thus connected, by way of silicon, with the root-nature, must also be able to be led upward through the plant. It must flow upward. There must be constant interaction between what is drawn in from the Cosmos by the silicon, and what takes place—forgive me!—in the "belly" up above; for by the latter process the "head" beneath must be supplied with what it needs. The "head" is supplied out of the Cosmos, but it must also be in mutual interaction with what is going on in the "belly," above the Earth's surface. In a word, that which pours down from the Cosmos and is caught up beneath the surface must be able to pour upward again. And for this purpose is the clayey substance in the soil. Everything in the nature of *clay* is in reality a means of transport, for the influences of cosmic entities within the soil, to carry them upward again from below.

When we pass on to practical matters, this knowledge will give us the necessary indications as to how we must deal with a clayey soil, or with a silicious soil, according as we have to plant it with one form of vegetation or another. First we must know what is really happening. However else clay may be described, however, else we may have to treat it so as to make it fertile—all that, no doubt, is most important in the second place, but the first thing is to know that clay is the carrier of the cosmic upward stream.

But this up-streaming of the cosmic influences is not all. There is also the other process which I may call the terrestrial or earthly —that process which is going on in the "belly" and which depends on a kind of external "digestion." For plant-growth, in effect, all that goes on through summer and winter in the air above the Earth is essentially a kind of digestion. All that is thus taking place through a kind of digestive process, must in its turn be drawn downward into the soil. Thus a true mutual interaction will arise with all the forces and fine homeopathic substances which are engendered by the water and air above the Earth. All this is drawn down into the soil by the greater or lesser *limestone* content of the soil. The limestone content of the soil itself, and the distribution of limestone substances in homeopathic dilution immediately above the soil—all this is there to carry into the soil the immediate terrestrial process.

In due time there will be a science of these things—not the mere scientific jargon of to-day—and it will then be possible to give exact indications. It will be known, for instance, that there is a very great difference between the warmth that is above the Earth's surface—that is to say, the warmth that is in the domain of Sun, Venus, Mercury and Moon—and that warmth which makes itself felt within the Earth; which is under the influence of Mars, Jupiter and Saturn. For the plant, we may describe the one kind as leaf-and-flower warmth, and the other as root warmth. These two warmths are essentially different, and in this sense, we may well call the warmth above the Earth *dead*, and that beneath the Earth's surface *living*.

The warmth beneath the Earth decidedly contains some inner principle of life. It is alive; moreover in winter it is most of all alive. If we human beings had to experience the warmth which works within the Earth, we should all grow dreadfully stupid, for to be clever we need to have *dead* warmth brought to our body. But the moment the warmth is drawn into the Earth by the limestone-content of the soil, or by other substantialities within the Earth—the moment any outer warmth passes over into inner warmth—it is changed into a certain condition of vitality, however delicate.

People to-day are well aware that there is a difference between the air above the soil and the air within, but they do not observe that there is also this difference between the *warmth* above and within. They know that the air beneath the surface contains more carbonic acid, and the air above, more oxygen, but again they do not know the reason. The reason is that the air too is permeated by a delicate vitality the moment it is absorbed and drawn into the Earth.

So it is both with the warmth and with the air; they take on a slightly living quality when they are received into the Earth. The opposite is true of the water and of the solid earthy element itself. They become still more dead inside the Earth than they are outside it. They lose something of their external life. Yet in this very process they become open to receive the most distant cosmic forces.

The mineral substances must emancipate themselves from what is working immediately above the surface of the Earth, if they wish to be exposed to the most distant cosmic forces. And in our cosmic age they can most easily do so—they can most easily emancipate themselves from the Earth's immediate neighbourhood and come under the influence of the most distant cosmic forces down inside the Earth—in the time between the 15th January and the 15th February; in this winter season. The time will come when such things are recognised as exact indications. This is the season when the strongest formative-forces of crystallisation, the strongest forces of form, can be developed for the mineral substances within the Earth. It is in the middle of the winter. The interior of the Earth then has the property of being least dependent on itself—on its own mineral masses; it comes under the influence of the crystal-forming forces that are there in the wide spaces of the Cosmos.

This then is the situation. Towards the end of January the mineral substances of the Earth have the greatest longing to become crystalline, and the deeper we go into the Earth, the more they have this longing to become purely crystalline within the "household of Nature." In relation to plant growth, what happens in the minerals at this time is most of all indifferent, or neutral. That is to say, the plants at this time are most left to themselves within the Earth; they are least exposed to the mineral substances. On the other hand, for a certain time before and after this period—and notably *before* it, when the minerals are, so to speak, just on the point of passing over into the crystalline element of form and shape—then they are of the greatest importance; they ray out the forces that are particularly important for plant-growth.

Thus we may say, approximately in the month of November-December, there is a point of time when that which is under the surface of the Earth becomes especially effective for plant-growth. The practical question is: "How can we really make use of this for the growth of plants?" The time will come when it is recognised, how very important it is to make use of these facts, so as to be able to direct the growth of plants. I will observe at once, if we are dealing with a soil which does not readily or of its own accord carry upward the influences which should be working upward in this winter season, then it is well to add a dose of clay to the soil. (I shall indicate the proper dose later on). We thereby prepare the soil to carry upward what, to begin with, is inside the Earth and make it effective for the growth of plants. I mean, the crystalline forces which we observe already when we look out over the crystallising snow. (The force of crystallisation, however, grows stronger and more intense the farther we go into the interior of the Earth). This crystallising force must therefore be carried upward at a time when it has not yet reached its culminating point—which it will only attain in January or February.

Thus we derive the most positive hints from knowledge which at first sight seems remote. We get indications that will really help us, where we should otherwise be experimenting in the dark.

Altogether, we should be clear that the whole domain of Agriculture—including what is beneath the surface of the Earth—represents an individuality, a living organism, living even in time. The life of the Earth is especially strong during the winter season, whereas in summer-time it tends in a certain sense to die.

Now for the tilling of the soil one important thing should above all be understood. I have often mentioned it among anthroposophists. It is this. We must know the conditions under which the cosmic spaces are able to pour their forces down into the earthly realm. To recognise these conditions, let us take our start from the seed-forming process. The seed, out of which the embryo develops, is usually regarded as a very complicated molecular structure, and scientists are especially anxious to understand it in its complex molecular structure. In simple molecules, they imagine, there is a simple structure; then it grows ever more complicated, till at last we get to the infinitely complex structure of the protein molecule.

With wonder and astonishment they stand before what they imagine as the complicated structure of the protein in the seed. For they conceive it as follows. They think the protein molecule must be extremely complicated; for after all, out of its complexity, the whole new organism will grow. The new organism, infinitely complex as it is, was already pre-figured in the embryonic condition of the seed. Therefore this microscopic or ultra-microscopic substance must also be infinitely complex in its structure.

To begin with, to a certain extent this is quite true. When the earthly protein is built up, the molecular structure is indeed raised to the highest complexity. But a new organism could never arise out of this complexity. The organism does not arise out of the seed in that way at all. That which develops as the seed, out of the mother-plant or mother-animal, does not by any means simply continue its existence in that which afterwards arises as the descendant plant or animal. That is not true. The truth is rather this:—

When the complexity of structure has been enhanced to the highest degree, it all disintegrates again, and eventually, where we first had the highest complexity attained within the Earth-domain, we now have a tiny realm of *chaos*. It all disintegrates, as we might say, into cosmic dust. Then, when the seed—having been raised to the highest complexity—has fallen asunder into cosmic dust and the tiny realm of chaos is there, then the entire surrounding Universe begins to work and stamps itself upon the seed, thus building up out of the tiny chaos that which can only be built in it by forces pouring in from the great Universe from all sides (Diagram No. 4). So in the seed we get an image of the Universe.

In every seed-formation, the earthly process of organisation is carried to the very end—to the point of chaos. Time and again, in the chaos of the seed the new organism is built up again out of the whole Universe. The parent organism has to play this part: through its affinity to a particular cosmic situation, it tends to bring the seed into that situation whereby the forces work from the right cosmic directions, so that a dandelion brings forth, not a barberry, but a dandelion in its turn.

That which is imaged in the single plant, is always the image of some cosmic constellation. Ever and again, it is built out of the Cosmos. Therefore, if ever we want to make the forces of the Cosmos effective in our earthly realm, we must drive the earthly as far as possible into a state of chaos. For plant-growth, Nature herself will see to it to some extent, that this is done. However, since every new organism is built out of the Cosmos, it is also necessary for us to preserve the cosmic process in the organism long enough—that is, until the seed-forming process occurs once more.

Say we plant the seed of some plant in the Earth. Here in this seed we have the stamp or impress of the whole Cosmos—from one cosmic aspect or another. The constellation takes effect in the seed; thereby it receives its special form. Now, the moment it is planted in the Earth-realm, the external forces of the Earth influence it very strongly, and it is permeated every moment with a longing to deny the cosmic process—that is to say, to grow hypertrophied, to grow out in all manner of directions. For that which is working above the Earth does not really want to preserve this form.

The seed must be driven to the state of chaos. On the other hand, when the first beginnings of the plant are unfolding out of the seed, and at the later stages also—over against the cosmic form which is living as the plant-form in the seed we need to bring the earthly element into the plant. We must bring the plant nearer to the Earth in its growth. And this we can only do by bringing into the life of the plant such life as is already present on the Earth. That is to say, we must bring into it life that has not yet reached the utterly chaotic state—life that has not yet gone forward to the stage of seed-formation—life, that is to say, which came to an end in the organisation of some plant before it reached the point of seed-formation.

In effect, we must bring into it such life as is already present on the Earth. In this respect, in districts which are well-favoured by fortune, a rich humus-formation comes very largely to man's assistance in "Nature's household." For in the last resort man can but sparingly replace by artificial means the fertility the Earth itself is able to achieve by natural humus-formation. To what is this humus-formation due? It is due to the fact that that which comes from the plant-life is absorbed by the whole Nature-process. To some extent, all life that has not yet reached the state of chaos rejects the cosmic influences. If such life is also made use of in the plant's growth, the effect is to hold fast in the plant what is essentially earthly. The cosmic process works only in the stream which passes upward once more to the seed-formation; while on the other hand the earthly process works in the unfolding of leaf, blossom and so on, and the cosmic only radiates its influences into all this.

We can trace the process quite exactly. Assume you have a plant growing upward from the root. At the end of the stem the little grain of seed is formed. The leaves and flowers spread themselves out. Now the earthly element in leaf and flower is the shape and form and the filling of earthly matter. The reason why a leaf or grain develops thick and strong—absorbs inner substantialities, and so on—the reason for this lies in all that which we bring to the plant by way of *earthly* life that has not yet reached the state of chaos. On the other hand, the seed which evolves its force right up the steam (in a vertical direction, not in the circling round)—the seed irradiates the leaf and blossom of the plant with the *force of the Cosmos*.

We can see this directly. Look at the green plant-leaves. (Diagram No. 3). The green leaves, in their form and thickness and in their greeness too, carry an earthly element, but they would not be green unless the cosmic force of the Sun were also living in them. And even more so when you come to the *coloured flower*; therein are living not only the cosmic forces of the Sun, but also the supplementary forces which the Sun-forces receive from the *distant planets*—Mars Jupiter and Saturn. In this way we must look at all plant growth. Then, when we contemplate the rose, in its red colour we shall see the forces of Mars. Or when we look at the yellow sunflower—it is not quite rightly so called, it is called so on account of its form; as to its yellowness it should really be named the Jupiter-flower. For the force of Jupiter, supplementing the cosmic force of the Sun, brings forth the white or yellow colour in the flowers. And when we approach the chicory (*Cichorium Intybus*), we shall divine in the bluish colour the influence of Saturn, supplementing that of the Sun. Thus we can recognise Mars in the red flower, Jupiter in the yellow or white, Saturn in the blue, while in the green leaf we see essentially the Sun itself. But that which thus shines out in the colouring of the flower works as a force most strongly in the root. For the forces that live and abound in the distant planets are working, as we have seen, down there below within the earthly soil.

It is so indeed. We must say to ourselves: Suppose we pull a plant out of the Earth. Down below we have the root. In the root there is the cosmic nature, whereas in the flower most of all there is the earthly, the cosmic being only present in the delicate quality of the colouring and shading. If on the other hand the earthly nature is to live strongly in the root, then it must shoot into form. For the plant always has its form from that which can arise within the earthly realm. That which expands the form is earthly. Thus if the root is ramified and much-divided, then, as in the flower's colouring the cosmic nature is working upward, so here the earthly nature is working downward. Therefore the cosmic roots are those that are more or less single in form, whereas in highly ramified roots we have a working of the earthly nature downward into the soil, just as in colour we have a working-upward of the cosmic nature into the flower.

The Sun-quality is in the midst between the two. The Sun-nature lives most of all in the green leaf, in the mutual interplay between the flower and the root and all that is between them. The Sun-quality is really that which is related, as a "diaphragm" (for so we called it in this picture) with the surface of the earth. The cosmic is associated with the interior of the Earth and works upward into the upper parts of the plant. The earthly, which is localised above the surface of the earth, works downward, being carried down into the plant with the help of the limestone element.

Observe those plants in which the limestone strongly draws the earthly nature downward into the roots. These are the plants whose roots shoot out in all directions with many ramifications, such, for instance, as the food fodder plants—I do not mean turnips or the like, but plants like sainfoin. Such things must be recognised in the form of the plant. To understand the plant, we must recognise the form of the plant and from the colour of the flower, the extent to which the cosmic and the earthly are working there.

Assume that by some means we cause the cosmic to be strongly retained—held up within the plant itself. Then it will not reveal itself to any great extent. It will not shoot out into blossom but will express itself in a stalk-like nature. Where, now, according to the indications we have given, does the cosmic nature live in the plant? It lives in the silicious element.

Look at the equisetum plant. It has this peculiarity: it draws the cosmic nature to itself; it permeates itself with the silicious nature. It contains no less than 90% of silicic acid. In equisetum the cosmic is present, so to speak, in very great excess, yet in such a way that it does not go upward and reveal itself in the flower but betrays its presence in the growth of the lower parts.

Or let us take another case. Suppose that we wish to hold back in the root-nature of a plant that which would otherwise tend upward through the stem and leaf. No doubt this is not so important in our present earthly epoch, for through various conditions we have already largely fixed the different species of plants. In former epochs—notably in primeval epochs—it was different. At that time it was still possible quite easily to transform one plant into another; hence it was very important to know these things. To-day too, it is important if we wish to find what conditions are favourable to one plant or another.

What do we then need to consider? How must we look at a plant when we desire the cosmic forces not to shoot upward into the blossoming and fruiting process but to remain below? Suppose we want the stem and leaf-formation to be held back in the root. What must we then do? We must put such a plant into a sandy soil, for in silicious soil the cosmic is held back; it is actually "caught." Take the potato, for example. With the potato this end must be attained. The blossoming process must be kept below. For the potato is a stem and leaf-formation down in the region of the root. The leaf and stem-forming process is held back, retained in the potato itself. The potato is not a root, it is a stem-formation held back. We must therefore bring it into a sandy soil. Otherwise we shall not succeed in having the cosmic force retained in the potato.

This, therefore, is the ABC for our judgment of plant-growth. We must always be able to say, what in the plant is cosmic, and what is terrestrial or earthly. How can we adapt the soil of the earth, by its special consistency, as it were to densify the cosmic and thereby hold it back more in the root and leaf? Or again, how can we thin it out so that it is drawn upward in a dilute condition, right up into the flowers, giving them colour—or into the fruit-forming process, permeating the fruit with a fine and delicate taste? For if you have apricots or plums with a fine taste—this taste, just like the colour of the flowers, is the cosmic quality which has been carried upward, right into the fruit. In the apple you are eating Jupiter, in the plum you are actually eating Saturn.

If mankind with their present state of knowledge were suddenly obliged to create, from the comparatively few plants of the primeval epoch of the Earth, the manifold variety of our present fruits and fruit-trees, they would not get very far. We should not get far if it were not for the fact that the forms of our different fruits are inherited. They were produced at a time when humanity had knowledge, out of primeval and instinctive widsom, how to create the different kinds of fruits from the primitive varieties that then existed. If we did not already possess the different kinds of fruit, handing them down by heredity—if we had to do it all over again with our present cleverness—we should not be very successful in creating the different kinds of fruit. Nowadays it is all done by blind experiment, there is no rational penetration into the process.

This must be re-discovered if we wish to go on working on the Earth at all. Extremely apt was the remark of our friend Stegemann to the effect that a decrease in the value of the products is observable. This decrease is indeed connected—like the transformation in the human soul itself—with the ending of Kali Yuga in the Universe during the last decades and in the decades that are now about to come. You may take my remark amiss or not, as you will. We stand face to face with a great change, even in the inner being of Nature. What has come down to us from ancient times—whatever it may be that we have handed down: natural talents, knowledge derived from Nature, and the like, even the traditional medicaments we still possess—all this is losing its value.

We must gain new knowledge in order to enter again into the whole Nature-relationship of these things. Mankind has no other choice. Either we must learn once more, in all domains of life—learn from the whole nexus of Nature and the Universe—or else we must see Nature and withal the life of Man himself degenerate and die. As in ancient times it was necessary for men to have knowledge entering into the inwardness of Nature, so do we now stand in need of such knowledge once again.

As I said just now, the man of to-day may know—though this knowledge too is very scanty—he may know how the air behaves in the interior of the Earth. But he knows practically nothing of how the *light* behaves in the interior of the Earth. He does not know that the silicious—that is, the cosmic—stone or rock or sand receives the light into the Earth and makes it effective there. Whereas that which stands nearer to the earthly-living nature, namely the humus, does not receive it; it does not make the light effective in the Earth. It therefore gives rise to a "light-less" working. Such things must be penetrated once more with clear understanding.

Now the plant-growth of the Earth is not all. To any given district of the Earth a specific animal life also belongs. For reasons which will presently be evident, we may for the moment leave man out, but we cannot neglect animal life. For this is the peculiar fact; the best—if I may call it so—cosmic qualitative analysis takes place of its own accord, in the life of a certain district of the Earth, overgrown as it is with plants, along with the animals in the same region. This is the peculiar fact—and I should be glad if my statements were tested, for if you subsequently test them you will certainly find them confirmed. This is the peculiar relation. If in any farm you have the right amount of horses, cows and other animals, these animals taken together will give just the amount of manure which you need for the farm itself, in order, as I said, to add something more to what has already turned into chaos.

Nay more, if you have the right number of cows, horses, pigs, etc., *severally*, the proportion of admixture in the manure will also be correct. This is due to the fact that the animals will eat the right measure of what is provided for them by the growth of plants. They eat the right quantity of what the Earth is able to provide. Hence in the course of their organic processes they bring forth just the amount of manure which needs to be given back again to the Earth.

This therefore is the case. We cannot carry it out absolutely, but in the ideal sense it is correct. If we are obliged to import any manure from outside the farm, properly speaking we should only use it as a remedy—as a medicament for a farm that has already grown ill. The farm is only healthy inasmuch as it provides its own manure from its own stock. Naturally, this will necessitate our developing a proper science of the number of animals of a given sort which we need for a given kind of farm. This need not cause any alarm. Such a science will arise in good time, as soon as we begin to have any knowledge again of the inner forces concerned.

In effect, what was said at the beginning of this lecture—describing that which is above the Earth's surface as a kind of belly, and that which is beneath as a kind of head-existence—is not complete unless we also understand the animal organism in this way. The animal organism lives in the whole complex of Nature's household. In form and colour and configuration, and in the structure and consistency of its substance from the front to the hinder parts, it is related to these influences. From the snout towards the heart, the Saturn, Jupiter and Mars influences are at work; in the heart itself the Sun, and behind the heart, towards the tail, the Venus, Mercury and Moon influences (Diagram No. 5). In this respect, those who are interested in these matters should develop their knowledge above all by learning to *read the form*. To be able to do this is of very great importance.

Go to a museum and look at the skeleton of any mammal, and go there with the consciousness that in the form and configuration of the head there is working above all the radiation of the Sun, the direct radiant influence of the Sun as it pours into the mouth. For reasons we shall yet discuss, the animal exposes itself to the Sun in a specific way. A lion exposes itself to the Sun differently from a horse. The forming of the head and that which immediately follows the head, depends on the way the animal is exposed to the Sun. Thus in the fore part of the animal we have the direct Sun-radiation, and as a consequence the forming and development of the head.

Now you will remember, the sunlight enters the sphere of the Earth in another way also. It is thrown back by the Moon. We have not only to do with the direct sunlight; we have also to do with the sunlight thrown back by the Moon. This sunlight thrown back by the Moon is quite ineffective when it shines on to the head of an animal. There it has no influence. (What I am now saying applies especially, however, to the embryo life). The light that is rayed back from the Moon develops its highest influence when it falls on the hinder parts of the animal. Look at the skeleton-formation of the hinder parts; observe its peculiar relation to the head-formation. Cultivate a sense of form to perceive this contrast—the attachment of the thighs, the forming of the outgoing parts of the digestive tract, in contrast to that which is formed as the opposite pole, from the head inward. There, in the fore and hinder parts of the animal, you have the true contrast of Sun and Moon.

Moreover you will find that the Sun-influence goes as far as the heart and stops short just before the heart. For the head and the blood-forming process, Mars, Jupiter and Saturn are at work. Then, from the heart backward, the Moon influence is supported by the Mercury and Venus forces. If therefore you turn the animal in this way and stand it on its head, with the head stuck into the Earth and the hinder parts upward—you have the position which the " agricultural individuality" has invisibly.

This will enable you to discover, from the form and figure of the animal, a definite relation between the manure, for example, which this animal provides, and the needs of the particular portion of the Earth, the plants of which the animal is eating. For you must know these things. You must know, for instance, that the cosmic influences which are effective in a plant rise upward from the interior of the Earth. They are led upward. Suppose a plant is especially rich in such cosmic influences. The animal which eats the plant will in its turn provide manure, out of its whole organism, on the basis of this fodder. Thereby it will provide the very manure which is most suited for the soil on which the plant is growing. Thus if you can read Nature's language of forms, you will perceive all that is needed by the "self-contained individuality" which a true farm or agricultural unit should be. Only the animal stock must also be included in it.

LECTURE THREE

KOBERWITZ,
11*th June*, 1924.

MY DEAR FRIENDS,

The earthly and cosmic forces, of which I have spoken, work in the farm through the substances of the Earth, needless to say. In the next lectures we shall pass on to various practical aspects, but before we can do so we must enter a little more precisely into the question: How do these forces work through the substances of the Earth? In the present lecture we shall consider Nature's activity quite generally speaking.

One of the most important questions in agriculture is that of the significance of *nitrogen*—its influence in all farm-production. This is generally recognised; nevertheless the question, what is the essence of nitrogen's activity, has fallen into great confusion nowadays. Wherever nitrogen is active, men only recognise, as it were, the last excrescence of its activities—the most superficial aspects in which it finds expression. They do not penetrate to the relationships of Nature wherein nitrogen is working, nor can they do so, so long as they remain within restricted spheres. We must look out into the wide spaces, into the wider aspects of Nature, and study the activities of nitrogen in the Universe as a whole. We might even say—and this indeed will presently emerge—that nitrogen as such does not play the first and foremost part in the life of plants. Nevertheless, to understand plant-life it is of the first importance for us to learn to know the part which nitrogen does play.

Nitrogen, as she works in the life of Nature, has so to speak four sisters, whose working we must learn to know at the same time if we would understand the functions and significance of nitrogen herslelf in Nature's so-called household. The four sisters of nitrogen are those that are united with her in plant and animal protein, in a way that is not yet clear to the outer science of to-day. I mean the four sisters, carbon. oxygen, hydrogen and sulphur.

To know the full significance of protein it will not suffice us to enumerate as its main ingredients hydrogen, oxygen, nitrogen and carbon. We must include another substance, of the profoundest importance for protein, and that is sulphur. Sulphur in protein is the very element which acts as mediator between the Spiritual that is spread throughout the Universe—the formative power of the Spiritual—and the physical.

Truly we may say, whoever would trace the tracks which the Spiritual marks out in the material world, must follow the activity of sulphur. Though this activity appears less obvious than that of other substances, nevertheless it is of great importance; for it is along the paths of sulphur that the Spiritual works into the physical domain of Nature. Sulphur is actually the carrier of the Spiritual. Hence the ancient name, "*sulphur*," which is closely akin to the name "*phosphorus*." The name is due to the fact that in olden time they recognised in the out-spreading, sun-filled light, the Spiritual itself as it spreads far and wide. Therefore they named "light-bearers" these substances—like sulphur and phosphorus—which have to do with the working of light into matter.

Seeing that sulphur's activity in the economy of Nature is so very fine and delicate, we shall, however, best approach it by first considering the four other sisters: carbon, hydrogen, nitrogen and oxygen. These we must first learn to understand; we shall see what they signify in the whole being of the Universe. The chemist of to-day knows little of these substances. He knows what they look like when he has them in his laboratory, but he knows practically nothing of their inner significance in the working of the Cosmos as a whole. The knowledge of modern chemistry about them is scarcely more than our knowledge of a man of whose outer form we caught a glimpse as we passed by him in the street—or maybe we took a snapshot of him, and with the help of the photograph we can now call him to mind. We must learn to know the deeper essence of these substances. What science does is scarcely more than to take snapshots of them with a camera. All that is said of them in scientific books and lectures is scarcely more than that.

Let us begin with *carbon*. (The application of these matters to plant-life will presently emerge). Carbon indeed has fallen in our time from a highly aristocratic status to a very plebeian one. Alas, how many other beings of the Universe have followed it along the same sad way! What do we see in carbon nowadays? That which we use, as coal, to heat our ovens! That which we use, as graphite, for our writing. True, we still assign an aristocratic value to one modification of carbon, namely diamond, but we have little opprotunity to value even that, for we can no longer afford to buy it!

What is known about carbon nowadays is very little when you consider its infinite significance in the Universe. The time is not so very long ago—only a few centuries—when this black fellow, carbon, was so highly esteemed as to be called by a very noble name. They called it the Stone of the Wise—the *Philosopher's Stone*. There has been much chatter as to what the "Stone of the Wise" may be. Very little has emerged from it. When the old alchemists and such people spoke of the Stone of the Wise, they meant carbon —in the various modifications in which it occurs. They held the name so secret and occult, only because if they had not done so, anyone and everyone would have possessed it—for it was only carbon. Why then was carbon the "Stone of the Wise"?

Here we can answer, with an idea from olden time, a point we need to understand again in our time when speaking about carbon. It is quite true, carbon occurs to-day in Nature in a broken, crumbled form, as coal or even graphite—broken and crumbled, owing to certain processes which it has undergone. How different it appears, however, when we perceive it in its living activity, passing through the human or animal body, or building up the plant-body out of its peculiar conditions. Then the amorphous, formless substance which we see as coal or carbon proves to be only the last excrescence—the corpse of that which coal or carbon truly is in Nature's household.

Carbon, in effect, is the bearer of all the creatively formative processes in Nature. Whatever in Nature is formed and shaped—be it the form of the plant persisting for a comparatively short time, or the eternally changing configuration of the animal body—carbon is everywhere the great plastician. It does not only carry in itself its black substantiality. Wherever we find it in full action and inner mobility, it bears within it the creative and formative cosmic pictures—the sublime cosmic Imaginations, out of which all that is formed in Nature must ultimately proceed.

There is a hidden plastic artist in carbon, and this plastician—building the manifold forms that are built up in Nature—makes use of sulphur in the process. Truly to see the carbon as it works in Nature, we must behold the Spirit-activity of the great Universe, moistening itself so-to-speak with sulphur, and working as a plastic artist—building with the help of carbon the more firm and well-defined form of the plant, or again, building the form in man, which passes away again the very moment it comes into being.

For it is thus that man is not plant, but man. He has the faculty, time and again to destroy the form as soon as it arises; for he excretes the carbon, bound to the oxygen, as *carbonic acid.* Carbon in the human body would form us too stiffly and firmly—it would stiffen our form like a palm. Carbon is constantly about to make us stiff and firm in this way, and for this very reason our breathing must constantly dismantle what the carbon builds. Our breathing tears the carbon out of its rigidity, unites it with the oxygen and carries it outward. So we are formed in the mobility which we as human beings need. In plants, the carbon is present in a very different way. To a certain degree it is fastened—even in annual plants—in firm configuration.

There is an old saying in respect of man: "Blood is a very special fluid"—and we can truly say: the human Ego, pulsating in the blood, finds there its physical expression. More accurately speaking, however, it is in the carbon—weaving and wielding, forming itself, dissolving the form again. It is on the paths of this carbon—moistened with sulphur—that that spiritual Being which we call the Ego of man moves through the blood. And as the human Ego—the essential Spirit of man—lives in the carbon, so in a manner of speaking the Ego of the Universe lives as the Spirit of the Universe—lives via the sulphur in the carbon as it forms itself and ever again dissolves the form.

In bygone epochs of Earth-evolution carbon alone was deposited or precipitated. Only at a later stage was there added to it, for example, the *limestone* nature which man makes use of to create something more solid as a basis and support—a solid scaffolding—for his existence. Precisely in order to enable what is living in the carbon to remain in perpetual movement, man creates an underlying framework in his limestone-bony skeleton. So does the animal, at any rate the higher animal. Thus, in his ever-mobile carbon-formative process, man lifts himself out of the merely mineral and rigid limestone-formation which the Earth possesses and which he too incorporates in order to have some solid Earth within him. For in the limestone form of the skeleton he *has* the solid Earth within him.

So you can have the following idea. Underlying all living things is a carbon-like scaffolding or framework—more or less rigid or fluctuating as the case may be—and along the paths of this framework the Spiritual moves through the World. Let me now make a drawing (purely diagrammatic) so that we have it before us visibly and graphically. (Diagram 6). I will here draw a scaffolding or framework such as the Spirit builds, working always with the help of sulphur. This, therefore, is either the ever-changing carbon—constantly moving in the sulphur, in its very fine dilution—or, as in plants, it is a carbon-frame-work more or less hard and fast, having become solidified, mingled with other ingredients.

Now whether it be man or any other living being, the living being must always be permeated by an *ethereal*—for the ethereal is the true bearer of life, as we have often emphasised. This, therefore, which represents the carbonaceous framework of a living entity, must in its turn be permeated by an ethereal. The latter will either stay still—holding fast to the beams of the framework—or it will also be involved in more or less fluctuating movement. In either case, the ethereal must be spread out, wherever the framework is. Once more, there must be something ethereal wherever the framework is. Now this ethereal, if it remained alone, could certainly not exist as such within our physical and earthly world. It would, so to speak, always slide through into the empty void. It could not hold what it must take hold of in the physical, earthly world, if it had not a physical carrier.

This, after all, is the peculiarity of all that we have on Earth: the Spiritual here must always have physical carriers. Then the materialists come, and take only the physical carrier into account, forgetting the Spiritual which it carries. And they are always in the right—for the first thing that meets us *is* the physical carrier. They only leave out of account that it is the *Spiritual* which must have a physical carrier everywhere.

What then is the physical carrier of that Spiritual which works in the ethereal? (For we may say, the ethereal represents the lowest kind of spiritual working). What is the physical carrier which is so permeated by the ethereal that the ethereal, moistened once more with sulphur, brings into it what it has to carry—not in formation this time, not in the building of the framework—but in eternal quickness and mobility into the midst of the framework? This physical element which with the help of sulphur carries the influences of life out of the universal ether into the physical, is none other than *oxygen.* I have sketched it here in green. If you regard it physically, it represents the oxygen. It is the weaving, vibrant and pulsating essence that moves along the paths of the oxygen. For the ethereal moves with the help of sulphur along the paths of oxygen.

Only now does the *breathing* process reveal its meaning. In breathing we absorb the oxygen. A modern materialist will only speak of oxygen such as he has in his retort when he accomplishes, say, an electrolysis of water. But in this oxygen the lowest of the supersensible, that is the ethereal, is living — unless indeed it has been killed or driven out, as it must be in the air we have around us. In the air of our breathing the living quality is killed, is driven out, for the living oxygen would make us faint. Whenever anything more highly living enters into us we become faint. Even an ordinary hypertrophy of growth—if it occurs at a place where it ought not to occur—will make us faint, nay even more than faint. If we were surrounded by living air in which the living oxygen were present, we should go about stunned and benumbed. The oxygen around us must be killed. Nevertheless, by virtue of its native essence it is the bearer of life—that is, of the ethereal. And it becomes the bearer of life the moment it escapes from the sphere of those tasks which are allotted to it inasmuch as it surrounds the human being outwardly, around the senses. As soon as it enters into us through our breathing it becomes alive again. Inside us it must be alive.

Circulating inside us, the oxygen is not the same as it is where it surrounds us externally. Within us, it is living oxygen, and in like manner it becomes living oxygen the moment it passes, from the atmosphere we breathe, into the soil of the Earth. Albeit it is not so highly living there as it is in us and in the animals, nevertheless, there too it becomes *living oxygen.* Oxygen under the earth is not the same as oxygen above the earth.

It is difficult to come to an understanding on these matters with the physicists and chemists, for—by the methods they apply—from the very outset the oxygen must always be drawn *out* of the earth-realm; hence they can only have dead oxygen before them. There is no other possibility for them. That is the fate of every science that only considers the physical. It can only understand the corpse. In reality, oxygen is the bearer of the living *ether*, and the living ether holds sway in it by using sulphur as its way of access.

But we must now go farther. I have placed two things side by side; on the one hand the carbon framework, wherein are manifested the workings of the highest spiritual essence which is accessible to us on Earth: the human Ego, or the cosmic spiritual Being which is working in the plants. Observe the human process: we have the breathing before us—the living oxygen as it occurs inside the human being, the living oxygen carrying the ether. And in the background we have the carbon-framework, which in the human being is in perpetual movement. These two must come together. The oxygen must somehow find its way along the paths mapped out by the framework. Wherever any line, or the like, is drawn by the carbon —by the spirit of the carbon—whether in man or anywhere in Nature there the ethereal oxygen-principle must somehow find its way. It must find access to the spiritual carbon-principle. How does it do so? Where is the mediator in this process?

The mediator is none other than *nitrogen*. Nitrogen guides the *life* into the *form* or configuration which is embodied in the carbon. Wherever nitrogen occurs, its task is to mediate between the life and the spiritual essence which to begin with is in the carbon-nature. Everywhere—in the animal kingdom and in the plant and even in the Earth—the bridge between carbon and oxygen is built by nitrogen. And the spirituality which—once again with the help of sulphur— is working thus in nitrogen, is that which we are wont to describe as the *astral*. It is the astral spirituality in the human astral body. It is the astral spirituality in the Earth's environment. For as you know, there too the astral is working—in the life of plants and animals, and so on.

Thus, spiritually speaking we have the astral placed between the oxygen and the carbon, and this astral impresses itself upon the physical by making use of nitrogen. Nitrogen enables it to work physically. Wherever nitrogen is, thither the astral extends. The ethereal principle of life would flow away everywhere like a cloud, it would take no account of the carbon-framework were it not for the nitrogen. The nitrogen has an immense power of attraction for the carbon-framework. Wherever the lines are traced and the paths mapped out in the carbon, thither the nitrogen carries the oxygen— thither the astral in the nitrogen drags the ethereal.

Nitrogen is for ever dragging the living to the spiritual principle. Therefore, in man, nitrogen is so essential to the life of the soul. For the soul itself is the mediator between the Spirit and the mere principle of life. Truly, this nitrogen is a most wonderful thing. If we could trace its paths in the human organism, we should perceive in it once more a complete human being. This "nitrogen-man" actually exists. If we could peal him out of the body he would be the finest ghost you could imagine. For the nitrogen-man imitates to perfection whatever is there in the solid human framework, while on the other hand it flows perpetually into the element of life.

Now you can see into the human breathing process. Through it man receives into himself the oxygen—that is, the ethereal life. Then comes the internal nitrogen, and carries the oxygen everywhere —wherever there is carbon, *i.e.*, wherever there is something formed and figured, albeit in everlasting change and movement. Thither the nitrogen carries the oxygen, so that it may fetch the carbon and get rid of it. Nitrogen is the real mediator, for the oxygen to be turned into carbonic acid and so to be breathed out.

This nitrogen surrounds us on all hands. As you know, we have around us only a small proportion of oxygen, which is the bearer of life, and a far larger proportion of nitrogen—the bearer of the astral spirit. By day we have great need of the oxygen, and by night too we need this oxygen in our environment. But we pay far less attention, whether by day or by night, to the nitrogen. We imagine that we are less in need of it—I mean now the nitrogen in the air we breathe. But it is precisely the nitrogen which has a spiritual relation to us. You might undertake the following experiment.

Put a human being in a given space filled with air, and then remove a small quantity of nitrogen from the air that fills the space, thus making the air around him slightly poorer in nitrogen than it is in normal life. If the experiment could be done carefully enough, you would convince yourselves that the nitrogen is immediately replaced. If not from without, then, as you could prove, it would be replaced from *within* the human being. He himself would have to give it off, in order to bring it back again into that quantitative condition to which, as nitrogen, it is accustomed. As human beings we must establish the right percentage-relationship between our whole inner nature and the nitrogen that surrounds us. It will not do for the nitrogen around us to be decreased. True, in a certain sense it would still suffice us. We do not actually need to breathe nitrogen. But for the spiritual relation, which is no less a reality, only the quantity of nitrogen to which we are accustomed in the air is right and proper. You see from this how strongly nitrogen plays over into the spiritual realm.

At this point I think you will have a true idea, of the necessity of nitrogen for the life of plants. The plant as it stands before us in the soil has only a physical and an ether-body; unlike the animal, it has not an astral body within it. Nevertheless, outside it the astral must be there on all hands. The plant would never blossom if the astral did not touch it from outside. Though it does not absorb it (as man and the animals do) nevertheless, the plant must be touched by the astral from outside. The astral is everywhere, and nitrogen itself—the bearer of the astral—is everywhere, moving about as a corpse in the air. But the moment it comes into the Earth, it is alive again. Just as the oxygen does, so too the nitrogen becomes alive; nay more it becomes sentient and sensitive inside the Earth. Strange as it may sound to the materialist madcaps of to-day, nitrogen not only becomes alive but *sensitive* inside the Earth; and this is of the greatest importance for agriculture. Nitrogen becomes the bearer of that mysterious sensitiveness which is poured out over the whole life of the Earth.

It is the nitrogen which senses whether there is the proper quantity of water in a given district of the Earth. If so, it has a sympathetic feeling. If there is too little water, it has a feeling of antipathy. It has a sympathetic feeling if the right plants are there for the given soil. In a word, nitrogen pours out over all things a kind of sensitive life. And above all, you will remember what I told you yesterday and in the previous lectures: how the planets, Saturn, Sun, Moon, etc., have an influence on the formation and life of plants. You might say, nobody knows of that! It is quite true, for ordinary life you can say so. Nobody knows! But the nitrogen that is everywhere present—the nitrogen knows very well indeed, and knows it quite correctly. Nitrogen is not unconscious of that which comes from the stars and works itself out in the life of plants, in the life of Earth. Nitrogen is the sensitive mediator, even as in our human nerves-and-senses system it is the nitrogen which mediates for our sensation. Nitrogen is verily the *bearer of sensation*. So you can penetrate into the intimate life of Nature if you can see the nitrogen everywhere, moving about like flowing, fluctuating feelings. We shall find the treatment of nitrogen, above all, infinitely important for the life of plants. These things we shall enter into later. Now, however, one thing more is necessary.

You have seen how there is a living interplay. On the one hand there is that which works out of the Spirit in the *carbon-principle*, taking on forms as of a scaffolding or framework. This is in constant interplay with what works out of the astral in the *nitrogen-principle*, permeating the framework with inner life, making it sentient. And in all this, life itself is working through the *oxygen-principle*. But these things can only work together in the earthly realm inasmuch as it is permeated by yet another principle, which for our physical world establishes the connection with the wide spaces of the Cosmos.

For earthly life it is impossible that the Earth should wander through the Cosmos as a solid thing, separate from the surrounding Universe. If the Earth did so, it would be like a man who lived on a farm but wanted to remain independent, leaving outside him all that is growing in the fields. If he is sensible, he will not do so! There are many things out in the fields to-day, which in the near future will be in the stomachs of this honoured company, and thence —in one way or another—it will find its way back again on to the fields. As human beings we cannot truly say that we are separate. We cannot sever ourselves. We are united with our surroundings—we belong to our environment. As my little finger belongs to me, so do the things that are around us naturally belong to the whole human being. There must be constant interchange of substance, and so it must be between the Earth—with all its creatures—and the entire Universe. All that is living in physical forms upon the Earth must eventually be led back again into the great Universe. It must be able to be purified and cleansed, so to speak, in the universal All. So now we have the following:—

To begin with, we have what I sketched before in blue (Diagram 6), the carbon-framework. Then there is that which you see here in the green—the ethereal, oxygen principle. And then—everywhere emerging from the oxygen, carried by nitrogen to all these lines—

there is that which develops as the astral, as the transition between the carbonaceous and the oxygen principle. I could show you everywhere, how the nitrogen carries into these blue lines what is indicated diagrammatically in the green.

But now, all that is thus developed in the living creature, structurally as in a fine and delicate design, must eventually be able to vanish again. It is not the Spirit that vanishes, but that which the Spirit has built into the carbon, drawing the life to itself out of the oxygen as it does so. This must be able once more to disappear. Not only in the sense that it vanishes on Earth; it must be able to vanish *into the Cosmos*, into the universal All.

This is achieved by a substance which is as nearly as possible akin to the physical and yet again as nearly akin to the spiritual—and that is *hydrogen.* Truly, in hydrogen—although it is itself the finest of physical elements—the physical flows outward, utterly broken and scattered, and carried once more by the sulphur out into the void, into the indistinguishable realms of the Cosmos.

We may describe the process thus: In all these structures, the Spiritual has become physical. There it is living in the body astrally, there it is living in its image, as the Spirit or the Ego—living in a physical way as Spirit transmuted into the physical. After a time, however, it no longer feels comfortable there. It wants to dissolve again. And now once more—moistening itself with sulphur—it needs a substance wherein it can take its leave of all structure and definition, and find its way outward into the undefined chaos of the universal All, where there is nothing more of this organisation or that.

Now the substance which is so near to the Spiritual on the one hand and to the substantial on the other, is hydrogen. Hydrogen carries out again into the far spaces of the Universe all that is *formed*, and *alive*, and *astral.* Hydrogen carries it upward and outward, till it becomes of such a nature that it can be received out of the Universe once more, as we described above. It is hydrogen which dissolves everything away.

So then we have these five substances. They, to begin with, represent what works and weaves in the living—and in the apparently dead, which after all is only transiently dead. Sulphur, carbon, hydrogen, oxygen, nitrogen: each of these materials is inwardly related to a specific spiritual principle. They are therefore very different from what our modern chemists would relate. Our chemists speak only of the corpses of the substances—not of the real substances, which we must rather learn to know as sentient and living entities, with the single exception of hydrogen. Precisely because hydrogen is apparently the thinnest element—with the least atomic weight—it is really the least spiritual of all.

And now I ask you to observe: When you meditate, what are you really doing? (I must insert this observation; I want you to see that these things are not conceived "out of the blue"). The Orientals used to meditate in their way; we in the mid-European West do it in our way. Our meditation is connected only *indirectly* with the breathing. We live and weave in concentration and meditation. However, all that we do when we devote ourselves to these exercises of the soul still has its bodily counterpart. Albeit this is delicate and subtle, nevertheless, however subtly, meditation somewhat modifies the regular course of our breathing, which as you know is connected so intimately with the life of man.

In meditating, we always retain in ourselves a little more carbon dioxide than we do in the normal process of waking consciousness. A little more carbon dioxide always remains behind in us. Thus we do not at once expel the full impetus of the carbonic acid, as we do in the everyday, bull-at-the-gate kind of life. We keep a little of it back. We do not drive the carbon dioxide with its full momentum out into the surrounding spaces, where the nitrogen is all around us. We keep it back a little.

If you knock up against something with your skull—if you knock against a table, for example—you will only be conscious of your own pain. If, however, you rub against it gently, you will be conscious of the surface of the table. So it is when you meditate. By and by you grow into a conscious living experience of the nitrogen all around you. Such is the real process in meditation. All becomes knowledge and perception—even that which is living in the nitrogen. And this nitrogen is a very clever fellow! He will inform you of what Mercury and Venus and the rest are doing. He knows it all, he really senses it. These things are based on absolutely real processes, and I shall presently touch on some of them in somewhat greater detail. This is the point where the Spiritual in our inner life begins to have a certain bearing on our work as farmers.

This is the point which has always awakened the keen interest of our dear friend Stegemann. I mean this working-together of the soul and Spirit in us, with all that is around us. It is not at all a bad thing if he who has farming to do can meditate. He thereby makes himself receptive to the revelations of nitrogen. He becomes more and more receptive to them. If we have made ourselves thus receptive to nitrogen's revelations, we shall presently conduct our farming in a very different style than before. We suddenly begin to *know* all kinds of things, all kinds of things emerge. All kinds of secrets that prevail in farm and farmyard—we suddenly begin to know them.

Nay more! I cannot repeat what I said here an hour ago, but in another way I may perhaps characterise it again. Think of a simple peasant-farmer, one whom your scholar will certainly not deem to be a learned man. There he is, walking out over his fields. The peasant is stupid—so the learned man will say. But in reality it is not true, for the simple reason that the peasant—forgive me, but it is so—is himself a *meditator*. Oh, it is very much that he meditates in the long winter nights! He does indeed acquire a kind of method

—a method of spiritual perception. Only he cannot express it. It suddenly emerges in him. We go through the fields, and all of a sudden the knowledge is there in us. We know it absolutely. Afterwards we put it to the test and find it confirmed. I in my youth, at least, when I lived among the peasant folk, could witness this again and again. It really is so, and from such things as these we must take our start once more. The merely intellectual life is not sufficient —it can never lead into these depths. We must begin again from such things. After all, the weaving life of Nature is very fine and delicate. We cannot sense it—it eludes our coarse-grained intellectual conceptions. Such is the mistake science has made in recent times. With coarse-grained, wide-meshed intellectual conceptions it tries to apprehend things that are far more finely woven.

All of these substances—sulphur, carbon, nitrogen, hydrogen—all are united together in protein. Now we are in a position to understand the process of seed-formation a little more fully than hitherto. Wherever carbon, hydrogen, nitrogen occur—in leaf or flower, calyx or root—everywhere they are bound to other substances in one form or another. They are dependent on these other substances; they are not independent. There are only two ways in which they can become independent: namely, on the one hand when the hydrogen carries them outward into the far spaces of the Universe—separates them all, carries them all away and merges them into an universal chaos; and on the other hand, when the hydrogen drives these fundamental substances of protein into the tiny seed-formation and makes them independent there, so that they become receptive to the inpouring forces of the Cosmos. In the tiny seed-formation there is chaos, and away in the far circumference there is chaos once more. Chaos in the seed must interact with chaos in the farthest circles of the Universe. Then the new being arises.

Now let us look how the action of these so-called substances—which in reality are bearers of the Spirit—comes about in Nature. You see, that which works even inside the human being as oxygen and nitrogen, behaves itself tolerably well. There in the human being the properties of oxygen and nitrogen are living. One only does not perceive them with ordinary science, for they are hidden to outward appearance. But the products of the carbon and hydrogen principles cannot behave quite so simply.

Take, to begin with, carbon. When the carbon, with its inherent activity, comes from the plant into the animal or human kingdom, it must first become mobile—in the transient stage at any rate. If it is then to present the firm and solid figure (man or animal), it must build on a more deep-seated scaffolding or framework. This is none other than the very deep-seated framework which is contained, not only in our bony skeleton with its *limestone*—nature, but also in the *silicious* element which we continually bear within us.

To a certain extent, the carbon in man and animal masks its native power of configuration. It finds a pillar of support in the configurative forces of limestone and silicon. Limestone gives it the earthly, silicon the cosmic formative power. Carbon, therefore, in man himself—and in the animal—does not declare itself exclusively competent, but seeks support in the formative activities of limestone and silicon.

Now we find limestone and silicon as the basis of plant growth too. Our need is to gain a knowledge of what the carbon develops throughout the process of digestion, breathing and circulation in man—in relation to the bony structure and the silicious structure. We must somehow evolve a knowledge of what is going on in there—inside the human being. We should be able to see it all, if we could somehow creep inside. We should see the carbonaceous formative activity raying out from the circulatory process into the calcium and silicon in man.

This is the kind of vision we must unfold when we look out over the surface of the Earth, covered as it is with plants and having beneath it the limestone and the silica—the calcium and silicon. We cannot look inside the human being; we must evolve the same knowledge by looking out over the Earth. There we behold the oxygen-nature caught up by the nitrogen and carried down into the carbon-nature. (The carbon itself, however, seeks support in the principles of calcium and silicon. We might also say, the process only passes *through* the carbon). That which is living in our environment—kindled to life in the oxygen—must be carried into the depths of the Earth, there to find support in the silica, working formatively in the calcium or limestone.

If we have any feeling or receptivity for these things, we can observe the process most wonderfully in the *papilionaceae* or *leguminosae*—in all those plants which are well known in farming as the nitrogen-collectors. They indeed have the function of drawing in the nitrogen, so to communicate it to that which is beneath them. Observe these leguminosae. We may truly say, down there in the Earth something is athirst for nitrogen; something is there that needs it, even as the lung of man needs oxygen. It is the limestone principle. Truly we may say, the limestone in the Earth is dependent on a kind of *nitrogen-inbreathing*, even as the human lung depends on the inbreathing of oxygen. These plants—the papilionaceae—represent something not unlike what takes place on our epithelial cells. By a kind of inbreathing process it finds its way down there.

Broadly speaking, the papilionaceae are the only plants of this kind. All other plants are akin, not to the inbreathing, but to the outbreathing process. Indeed, the entire organism of the plant-world is dissolved into two when we contemplate it in relation to nitrogen. Observe it as a kind of nitrogen-breathing, and the entire organism of the plant-world is thus dissolved. On the one hand, where we encounter any species of papilionaceae, we are observing as it were the paths of the breathing, and where we find any other plants, there we are looking at the remaining organs, which breathe in a far more hidden way and have indeed other specific functions. We must learn to regard the plant-world in this way. Every plant-species must appear to us, placed in the total organism of the plant-world, like the single human organs in the total organism of man. We must regard the several plants as parts of a totality. Look on the matter in this way, and we shall perceive the great significance of the papilionaceae. It is no doubt already known, but we must also recognise the spiritual foundations of these things. Otherwise the danger is very great that in the near future, when still more of the old will be lost, men will adopt false paths in the application of the new.

Observe how the papilionaceae work. They all have the tendency to retain, to some extent in the region of the leaf-like nature, the *fruiting* process which in the other plants goes farther upward. They have a tendency to fruit even before the flowering process. You can see this everywhere in the papilionaceae; they tend to fruit even before they come to flower. It is due to the fact that they retain far nearer to the Earth that which expresses itself in the nitrogen nature. Indeed, as you know, they actually carry the nitrogen-nature into the soil.

Therefore, in these plants, everything that belongs to nitrogen lives far more nearly inclined to the Earth than in the other plants, where it evolves at a greater distance from the Earth. See how they tend to colour their leaves, not with the ordinary green, but often with a darker shade. Observe too how the fruit, properly speaking, tends to be stunted. The seeds, for instance, only retain their germinating power for a short time, after which they lose it.

In effect, these plants are so organised as to bring to expression, most of all, what the plant-world receives from the winter—not what it has from the summer. Hence, one would say, there is always a tendency in these plants to wait for the winter. With all that they evolve, they tend to wait for the winter. Their growth is retarded when they find a sufficiency of what they need—*i.e.*, of the nitrogen of the air, which in their own way they can carry downward.

In such ways as these we can look into the life and growth of all that goes on in and above the surface of the soil. Now you must also include this fact: the limestone-nature has in it a wonderful kinship to the world of human cravings. See how it all becomes organic and alive! Take the chalk or limestone when it is still in the form of its element—as calcium. Then indeed it gives no rest at all. It wants to feel and fill itself at all costs; it wants to become quicklime—that is, to unite its calcium with oxygen. Even then it is not satisfied, but craves for all sorts of things—wants to absorb all manner of metallic acids, or even bitumen which is scarcely mineral at all. It wants to draw everything to itself. Down there in the ground it unfolds a regular craving-nature.

He who is sensitive will feel this difference, as against a certain other substance. Limestone sucks us out. We have the distinct feeling: wherever the limestone principle extends, there is something that reveals a thorough craving nature. It draws the very plant-life to itself. In effect, all that the limestone desires to have, lives in the plant-nature. Time and again, this must be wrested away from it. How so? By the most aristocratic principle—that which desires nothing for itself. There *is* such a principle, which wants for nothing more but rests content in itself. That is the *silica*-nature. It has indeed come to rest in itself.

If men believe that they can only see the silica where it has hard mineral outline, they are mistaken. In homeopathic proportions, the silicious principle is everywhere around us; moreover it rests in itself—it makes no claims. Limestone claims everything; the silicon principle claims nothing for itself. It is like our own sense-organs. They too do not perceive themselves, but that which is outside them. The *silica*-nature is the universal *sense* within the earthly realm, the *limestone*-nature is the universal *craving*; and the *clay* mediates between the two. Clay stands rather nearer to the silicious nature, but it still mediates towards the limestone.

These things we ought at length to see quite clearly; then we shall gain a kind of sensitive cognition. Once more we ought to feel the chalk or limestone as the kernel-of-desire. Limestone is the fellow who would like to snatch at everything for himself. Silica, on the other hand, we should feel as the very superior gentleman who wrests away all that can be wrested from the clutches of the limestone, carries it into the atmosphere, and so unfolds the forms of plants. This aristocratic gentleman, silica, lives either in the ramparts of his castle—as in the equisetum plant — or else distributed in very fine degree, sometimes indeed in highly homeopathic doses. And he contrives to tear away what *must* be torn away from the limestone.

Here once more you see how we encounter Nature's most wonderfully intimate workings. Carbon is the true form-creator in all plants; carbon it is that forms the framework or scaffolding. But in the course of earthly evolution this was made difficult for carbon. It *could* indeed form the plants if it only had water beneath it. Then it would be equal to the task. But now the limestone is there beneath it, and the limestone disturbs it. Therefore it allies itself to silica. Silica and carbon together—in union with clay, once more —create the forms. They do so in alliance because the resistance of the limestone-nature must be overcome.

How then does the plant itself live in the midst of this process? Down there below, the limestone-principle tries to get hold of it with tentacles and clutches, while up above the silica would tend to make it very fine, slender and fibrous—like the aquatic plants. But in the midst—giving rise to our actual plant forms—there is the carbon, which orders all these things. And as our *astral body* brings about an inner order between our *Ego* and our *ether body*, so does the nitrogen work in between, as the astral.

All this we must learn to understand. We must perceive how the nitrogen is there at work, in between the lime—the clay—and the silicious—natures—in between all that the limestone of itself would constantly drag downward, and the silica of itself would contantly ray upward. Here then the question arises, what is the proper way to bring the nitrogen-nature into the world of plants? We shall deal with this question to-morrow, and so find our way to the various forms of manuring.

DISCUSSIONS

ADDRESS BY DR. RUDOLF STEINER

KOBERWITZ,
11th June, 1924.

MY DEAR FRIENDS,

Allow me in the first place to express my deep satisfaction that this Experimental Circle has been created as suggested by Count Keyserlingk, and extended to include all those concerned with agriculture who are now present for the first time at such a meeting. In point of time, the foundation has come about as follows. To begin with, Herr Stegemann, in response to several requests, communicated some of the things which he and I had discussed together in recent years concerning the various guiding lines in agriculture, which he himself has tested in one way or another in his very praiseworthy endeavours on his own farm. Thence there arose a discussion between him and our good friend Count Keyserlingk, leading in the first place to a consultation during which the resolution which has to-day been read out was drafted.

As a result of this we have come together here to-day. It is deeply satisfying that a number of persons have now found themselves together who will be the bearers, so to speak, of the experiments which will follow the guiding lines (for to begin with they can only be guiding lines) which I have given you in these lectures. These persons will now make experiments in confirmation of these guiding lines, and demonstrate how well they can be used in practice.

At such a moment, however, when so good a beginning has been made, we should also be careful to turn to good account the experiences we have had in the past with our attempts in other domains in the Anthroposophical Movement. Above all, we should avoid the mistakes which only became evident during the years when from the central anthroposophical work—if I may so describe it—we went on to other work which lay more at the periphery. I mean when we began to introduce what Anthroposophical Science must and can be for the several domains of life.

For the work which this Agricultural Circle has before it, it will not be without interest to hear the kind of experiences we have had in introducing Anthroposophical Science, for example, into the scientific life in general. As a general rule, when it came to this point, those who had hitherto administered the central anthroposophical life with real inner faithfulness and devotion in their own way, and those who stood more at the periphery and wanted to apply it to a particular domain of life, did not as a rule confront one another with full mutual understanding.

We experienced it only too well, especially in working with our scientific Research Institutes. There on the one side are the anthroposophists who find their full life in the heart of Anthroposophia itself—in Anthroposophical Science as a world-conception, a content of life which they may even have carried through the world with strong and deep feeling, every moment of their lives. There are the anthroposophists who live Anthroposophia and love it, making it the content of their lives. Generally, though not always, they have the idea that something important has been done when one has gained, here or there, one more adherent, or perhaps several more adherents, for the anthroposophical movement. When they work outwardly at all, their idea seems to be—you will forgive the expression—that people must somehow be able to be won over "by the scruff of the neck." Imagine, for example, a University professor in some branch of Natural Science. Placed as he is in the very centre of the scientific work on which he is engaged, he ought none the less to be able to be won over there and then—so they imagine.

Such anthroposophists, with all their love and good-will, naturally imagine that we should also be able to get hold of the farmer there and then—to get him too "by the scruff of the neck," so to speak, from one day to another, into the anthroposophical life—to get him in "lock, stock and barrel" with the land and all that is comprised with it, with all the products which his farm sends out into the world. So do the "central anthroposophists" imagine. They are of course in error. And although many of them say that they are faithful followers of mine, often, alas! though it is true enough that they are faithful in their inner feeling, they none the less turn a deaf ear to what I have to say in decisive moments. They do not hear it when I say, for instance, that it is utterly naive to imagine that you can win over to Anthroposophical Science some professor or scientist or scholar from one day to the next and without more ado. Of course you cannot. Such a man would have to break with twenty or thirty years of his past life and work, and to do so, he would have to leave an abyss behind him. These things must be faced as they exist in real life. Anthroposophists often imagine that life consists merely in thought. It does not consist in mere thought. I am obliged to say these things, hoping that they may fall upon the right soil.

On the other hand, there are those who out of good and faithful hearts want to unite some special sphere of life with Anthroposophia—some branch of science, for example. They also did not make things quite clear to themselves when they became workers in Spiritual Science. Again and again they set out with the mistaken opinion that we must do these things as they have hitherto been done in Science; that we must proceed precisely in the same way. For instance, there are a number of very good and devoted anthroposophists working with us in Medicine (with regard to what I shall now say, Dr. Wegman is an absolute exception; she always saw quite clearly the necessity prevailing in our Society). But a number of them always seemed to believe that the doctor must now apply what proceeds from anthroposophical therapy in the same medical style and manner to which he has hitherto been accustomed.

What do we then experience? Here it is not so much a question of spreading the central teachings of Spiritual Science; here it is more a question of spreading the anthroposophical life into the world. What did we experience? The other people said "Well, we have done that kind of thing before; we are the experts in that line. That is a thing we can thoroughly grasp with our own methods; we can judge of it without any doubt or difficulty. And yet, what these anthroposophists are bringing forward is quite contrary to what we have hitherto found by our methods." Then they declared that the things we say and do are wrong.

We had this experience: If our friends tried to imitate the outer scientists, the latter replied that they could do far better. And in such cases it was undeniable; they *can* in fact apply their methods better, if only for the reason that in the science of the last few years the methods have been swallowing up the science! The sciences of to-day seem to have nothing left but methods. They no longer set out on the objective problems; they have been eaten up by their own methods. To-day therefore, you can have scientific researches without any substance to them whatever.

And we have had this experience: Scientists who had the most excellent command of their own methods became violently angry when anthroposophists came forward and did nothing else but make use of these methods. What does this prove? In spite of all the pretty things that we could do in this way, in spite of the splendid researches that are being done in the Biological Institute, the one thing that emerged was that the other scientists grew wild with anger when our scientists spoke in their lectures on the basis of the very same methods. They were wild with anger, because they only heard again the things they were accustomed to in their own grooves of thought.

But we also had another important experience, namely this: A few of our scientists at last bestirred themselves, and departed to some extent from their old custom of imitating the others. But they only did it half and half. They did it in this way: In the first part of their lectures they would be thoroughly scientific; in the first part of their explanations they would apply all the methods of science, "comme il faut." Then the audience grew very angry. "Why do they come, clumsily meddling in our affairs? Impertinent fellows, these anthroposophists, meddling in their dilettante way with *our* science!"

Then, in the second part of their lectures, our speakers would pass on to the essential life—no longer elaborated in the old way, but derived as anthroposophical content from realms beyond the Earth. And the same people who had previously been angry became exceedingly attentive, hungry to hear more. Then they began to catch fire! They liked the Spiritual Science well enough, but they could not abide (and what is more, as I myself admitted, *rightly* not), what had been patched together as a confused "mixtum compositum" of Spiritual Science and Science. We cannot make progress on such lines.

I therefore welcome with joy what has now arisen out of Count Keyserlingk's initiative, namely that the professional circle of farmers will now unite on the basis of what we have founded in Dornach—the Natural Science Section. This Section, like all the other things that are now coming before us, is a result of the Christmas Foundation Meeting. From Dornach, in good time, will go out what is intended. There we shall find, out of the heart of Anthroposophia itself, scientific researches and methods of the greatest exactitude.

Only, of course, I cannot agree with Count Keyserlingk's remark that the professional farmers' circle should only be an executive organ. From Dornach, you will soon be convinced, guiding lines and indications will go out which will call for everyone at his post to be a fully independent fellow-worker, provided only that he wishes to work with us. Nay more, as will emerge at the end of my lectures (for I shall have to give the first guiding lines for this work at the close of the present lectures) the foundation for the beginning of our work at Dornach will in the first place have to come from you. The guiding lines we shall have to give will be such that *we* can only begin on the basis of the answers we receive from you.

From the beginning, therefore, we shall need most active fellow-workers—no mere executive organs. To mention only one thing, which has been a subject of frequent discussions in these days between Count Keyserlingk and myself—an agricultural estate is always an individuality, in the sense that it is never the same as any other. The climate, the conditions of the soil, provide the very first basis for the individuality of a farm. A farming estate in Silesia is not like one in Thuringia, or in South Germany. They are real individualities.

Now, above all in Spiritual Science, vague generalities and abstractions are of no value, least of all when we wish to take a hand in practical life. What is the value of speaking only in vague and general terms of such a practical matter as a farm is? We must always bear in mind the concrete things; then we can understand what has to be applied. Just as the most varied expressions are composed of the twenty-six letters of the alphabet, so you will have to deal with what has been given in these lectures. What you are seeking will first have to be composed from the indications given in these lectures —as words are composed from the letters of the alphabet. If on the basis of our sixty members we wish to speak of practical questions, our task, after all, will be to find the practical indications and foundations of work for those sixty individual farmers.

The first thing will be to gather up what we already know. Then our first series of experiments will follow, and we shall work in a really practical way. We therefore need the most active members. That is what we need in the Anthroposophical Society as a whole—good, practical people who will not depart from the principle that practical life, after all, calls forth something that cannot be made real from one day to the next. If those whom I have called the "central anthroposophists" believe that a professor, farmer or doctor—who has been immersed for decades past in a certain milieu and atmosphere—can accept anthroposophical convictions from one day to the next, they are greatly mistaken.

The fact will emerge quickly enough in agriculture! The farming anthroposophist no doubt, if he is idealistic enough, can go over entirely to the anthrospophical way of working —say, between his twenty-ninth and his thirtieth year—even with the work on his farm. But will his fields do likewise? Will the whole organisation of the farm do likewise? Will those who have to mediate between him and the consumer do likewise—and so on and so on? You cannot make them all anthroposophists at once—from your twenty-ninth to your thirtieth year. And when you begin to see that you cannot do so, it is then that you lose heart. That is the point, my dear friends—*do not lose heart*; know that it is not the momentary success that matters; it is the working on and on with *iron perseverance.*

One man can do more, another less. In the last resort, paradoxical as it may sound, you will be able to do more, the more you restrict yourself in regard to the area of land which you begin to cultivate in our ways. After all, if you go wrong on a small area of land, you will not be spoiling so much as you would on a larger area. Moreover, such improvements as result from our anthroposophical methods will then be able to appear very rapidly, for you will not have much to alter. The inherent efficiency of the methods will be proved more easily than on a large estate. In so practical a sphere as farming these things must come about by mutual agreement if our Circle is to be successful. Indeed, it is very strange—with all good humour and without irony, for one enjoyed it—there has been much talk in these days as to the differences that arose in the first meeting between the Count and Herr Stegemann. Such things bring with them a certain colouring; indeed, I almost thought I should have to consider whether the anthroposophical "Vorstand," or some one else, should not be asked to be present every evening to bring the warring elements together.

By and by however, I came to quite a different conclusion; namely, that what is here making itself felt is the foundation of a rather intimate mutual tolerance among farmers—an intimate "live and let live" among fellow-farmers. They only have a rough exterior. As a matter of fact the farmer, more than many other people, needs to protect his own skin. It can easily happen that people start interfering with things which he alone understands. And at rock bottom you will discover in him a certain sweet tolerance. All these things must be truly felt, and I only make these observations now because I think it necessary to begin on a right basis from the outset.

Therefore I think I may once again express my deep satisfaction at what has been done by you here. I believe we have truly taken into account the experiences of the Anthroposophical Society. What has now been begun will be a thing of great blessing, and Dornach will not fail to work vigorously with those who wish to be with us as active fellow-workers in this cause.

We can only be glad, that what is now being done in Koberwitz has been thus introduced. And if Count Keyserlingk so frequently refers to the burden I took upon myself in coming here, I for my part would answer—though not in order to call up any more discussion:– What trouble have I had? I had only to travel here, and am here under the best and most beautiful conditions. All the unpleasant tasks are undertaken by others; I only have to speak every day, though I confess I stood before these lectures with a certain awe—for they enter into a new domain. My trouble after all, was not so great. But when I see all the trouble to which Count Keyserlingk and his whole household have been put—when I see those who have come here—then I must say, for so it seems to me, that all the countless things that had to be done by those who have helped to enable us to be together here, tower above what I have had to do, who have simply sat down in the middle of it all when all was ready.

In this, then, I cannot agree with the Count. Whatever appreciation or gratitude you feel for the fact that this Agricultural Course has been achieved, I must ask you to direct your gratitude to him, remembering above all that if he had not thought and pondered with such iron strength, and sent his representative to Dornach, never relinquishing his purpose—then, considering the many things that have to be done from Dornach, it is scarcely likely that this Course in the farthest Eastern corner of the country could have been given.

Hence I do not at all agree that your feelings of gratitude should be expended on me, for they belong in the fullest sense to Count Keyserlingk and to his House.

That is what I wished to interpolate in the discussion.

For the moment, there is not much more to be said—only this. We in Dornach shall need, from everyone who wishes to work with us in the Circle, a description of what he has beneath his soil, and what he has above it, and how the two are working together. If our indications are to be of use to you, we must know exactly what the things are like, to which these indications refer. You from your practical work will know far better than we can know in Dornach, what is the nature of your soil, what kind of woodland there is and how much, and so on; what has been grown on the farm in the last few years, and what the yield has been. We must know all these things, which, after all, every farmer must know for himself if he wants to run his farm in an intelligent way—with "peasant wit."

These are the first indications we shall need: what is there on your farm, and what your experiences have been. That is quickly told. As to how these things are to be put together, that will emerge during the further course of the conference. Fresh points of view will be given which may help some of you to grasp the real connections between what the soil yields and what the soil itself is, with all that surrounds it. With these words I think I have adequately characterised the form which Count Keyserlingk wished the members of the Circle to fill in. As to the kind and friendly words which the Count has once again spoken to us all, with his fine-feeling distinction between "farmers" and "scientists," as though all the farmers were in the Circle and all the scientists at Dornach—this also cannot and must not remain so. We shall have to grow far more together; in Dornach itself, as much as possible of the peasant-farmer must prevail, in spite of our being "scientific." Moreover, the science that shall come from Dornach must be such as will seem good and evident to the most conservative, "thick-headed" farmer.

I hope it was only a kind of friendliness when Count Keyserlingk said that he did not understand me—a special kind of friendliness. For I am sure we shall soon grow together like twins—Dornach and the Circle. In the end he called me a "Grossbauer," that is, a yeoman farmer—thereby already showing that he too has a feeling that we *can* grow together. All the same, I cannot be addressed as such merely on the strength of the little initial attempt I made in stirring the manure—a task to which I had to give myself just before I came here. (Indeed it had to be continued, for I could not go on stirring long enough. You have to stir for a long time; I could only begin to stir, then someone else had to continue).

These are small matters, but it was not out of this that I originally came. I grew up entirely out of the peasant folk, and in my spirit I have always remained there—I indicated this in my autobiography. Though it was not on a large farming estate such as you have here, in a smaller domain I myself planted potatoes, and though I did not breed horses, at any rate I helped to breed pigs. And in the farmyard of our immediate neighbourhood I lent a hand with the cattle. These things were absolutely near my life for a long time; I took part in them most actively. Thus I am at any rate lovingly devoted to farming, for I grew up in the midst of it myself, and there is far more of that in me than the little bit of "stirring the manure" just now.

Perhaps I may also declare myself not quite in agreement with another matter at this point. As I look back on my own life, I must say that the most valuable farmer is not the large farmer, but the small peasant farmer who himself as a little boy worked on the farm. And if this is to be realised on a larger scale—translated into scientific terms—then it will truly have to grow "out of the skull of a peasant," as they say in Lower Austria. In my life this will serve me far more than anything I have subsequently undertaken.

Therefore, I beg you to regard me as the small peasant farmer who has conceived a real love for farming; one who remembers his small peasant farm and who thereby, perhaps, can understand what lives in the peasantry, in the farmers and yeomen of our agricultural life. They will be well understood at Dornach; of that you may rest assured. For I have always had the opinion (this was not meant ironically, though it seems to have been misunderstood) I have always had the opinion that their alleged stupidity or foolishness is "wisdom before God," that is to say, before the Spirit. I have always considered what the peasants and farmers thought about their things far wiser than what the scientists were thinking. I have invariably found it wiser, and I do so to-day. Far rather would I listen to what is said of his own experiences in a chance conversation, by one who works directly on the soil, than to all the Ahrimanic statistics that issue from our learned science. I have always been glad when I could listen to such things, for I have always found them extremely wise, while, as to science—in its practical effects and conduct I have found it very stupid. This is what we at Dornach are striving for, and this will make our science wise—will make it wise precisely through the so-called "peasant stupdity." We shall take pains at Dornach to carry a little of this peasant stupidity into our science. Then this stupidity will become —"wisdom before God."

Let us then work together in this way; it will be a genuinely conservative, yet at the same time a most radical and progressive beginning. And it will always be a beautiful memory to me if this Course becomes the starting point for carrying some of the real and genuine "peasant wit" into the methods of science. I must not say that these methods have become stupid, for that would not be courteous, but they have certainly become dead.

Dr. Wachsmuth has also set aside this deadened science, and has called for a living science which must first be fertilised by true "peasant wisdom." Let us then grow together thus like good Siamese Twins—Dornach and the Circle. It is said of twins that they have a common feeling and a common thinking. Let us then have this common feeling and thinking; then we shall go forward in the best way in our domain.

LECTURE FOUR

KOBERWITZ,
12th June, 1924.

MY DEAR FRIENDS,

You have now seen what is essential in the discovery of spiritual-scientific methods for Agriculture, as it is for other spheres of life. Nature and the working of the Spirit throughout Nature must be recognised on a large scale, in an all-embracing sphere. Materialistic science has tended more and more to the investigation of minute, restricted spheres. True, this is not quite so bad in Agriculture; here they do not always go on at once to the very minute—the microscopically small, with which they are wont to deal in other sciences. Nevertheless, here too they deal with narrow spheres of activity, or rather, with conclusions which they feel able to draw from the investigation of narrow and restricted spheres. But the world in which man and the other earthly creatures live cannot possibly be judged from such restricted aspects.

To deal with the realities of Agriculture as the customary science of to-day would do, is as though one would try to recognise the full being of man, starting from the little finger or from the lobe of the ear and trying to construct from thence the total human being. Here again we must first establish a genuine science—a science that looks to the great cosmic relationships. This is most necessary nowadays. Think how the customary science of to-day, or yesterday, has to correct itself. You need only remember the absurdities that prevailed not long ago in the science of human nutrition, for example. The statements were "absolutely scientific"—"scientifically proven"—and indeed, if one concentrated on the limited aspects which were brought forward, one could not make objection to the proofs. It was scientifically proven that a human being of average weight (eleven to twelve stone) requires about four-and-one-quarter ozs. of protein a day for adequate nourishment. It was, so to speak, an established fact of science. And yet, to-day no man of science believes in this proposition. Science has corrected itself in the meantime. To-day as everybody knows, four-and-one-quarter oz. of albuminous food are not only unnecessary but positively harmful, and a man will remain most healthy if he only eats one-and-three-quarter oz. a day.

In this instance, science has corrected itself, and it is well-known that if superflous protein is consumed, it will create by-products in the intestines—by-products which have a toxic effect. Examine not only the period of life in which the protein is taken, but the whole life of the human being, and you recognise that the arterial sclerosis of old age is largely due to the toxic effect of superfluous protein. In this way scientific investigations are often erroneous—in relation to man, for instance—inasmuch as they only deal with the given moment. A normal human life lasts longer than ten years, and the harmful effects of the seemingly good causes which they mistakenly strive to produce, often do not emerge for a long time. Spiritual Science will not fall so easily into such errors.

I do not wish to join in the facile criticisms which are so frequently made against orthodox science because it has to correct itself as in this instance. One can understand that it cannot be otherwise. No less facile, on the other hand, are the attacks that are made on Spiritual Science when it begins to enter into practical life, recognising as it does the wider connections. For in these larger relationships of life, Spiritual Science is impressed by those substances and forces which go out eventually into the spiritual realm. It does not merely recognise the coarse material forces and substantialities.

This applies also to Agriculture, and notably when we come to the question of *manuring*. The very way the words are often put by scientists when they come to the manuring question, shows how little idea they really have of what manuring signifies in the economy of Nature. How often do we hear the phrase: "Manure contains the necessary foodstuffs for the plants." I spoke these introductory sentences just now—referring to the nourishment of man—not without reason. I wanted to show you how science has had to correct itself in this instance, notably in the most recent period. Why has it to correct itself? Because it takes its start from an altogether false idea of nutrition—whether of man or of any other living creature.

Do not be angry with me for saying these things so openly and clearly. The idea used to be that the essential thing in human nutrition is what a man daily consumes. Undoubtedly, our daily food *is* important. But the greater part of what we daily eat is not there to be received *as substance* into the body—to be deposited in the body substantially. By far the greater part is there to give the body the *forces* which it contains, and so to call forth in the body inner mobility, activity. The greater part of what man thus receives into himself is cast out again.

Therefore the important question in the metabolic process is not the proportion of weights, but it is this: Are the foodstuffs providing us with the proper living quality of forces? We need these living forces, for example, when we walk or when we work—nay, when we only move our arms about. What the body needs, on the other hand, so as to deposit substances in itself—to provide itself with substances (which are expelled again every seven or eight years as the substance of the body is renewed)—this, for the most part, is received through the sense-organs, the skin and the breathing. Whatever the body has to receive and deposit in itself as actual *substance*—this it is constantly receiving in exceedingly minute doses, in a highly diluted state. It is only *in* the body that it becomes condensed. The body receives it from the air and thereupon hardens and condenses it, till in the nails and hair for instance it has to be cut off.

It is completely wrong to set up the formula: "Food received—Passage through the body—Wearing-away of nails and skin, and the like." The true formula is thus: "Breathing, or reception of substances in an even finer state through the sense-organs (even the eyes)—Passage through the organism—Excretion in the widest sense." On the other hand, what we receive through our stomach is important by virtue of its inherent life and mobility—as of a fuel. It is important inasmuch as it introduces the necessary forces for the *will* which is at work in the body. This is the truth—the simple result of spiritual research.

Over against this truth, it is heart-rending to see the ideas of modern science proclaiming the exact opposite. I say heart-rending, because we must admit, it is very difficult to come to terms at all with this science of to-day, even in the most essential questions. Yet somehow we *must* come to terms with it. For in practical life, the science of to-day would very soon lead into an absolute blind-alley. While it pursues its present path it is simply incapable of understanding certain matters even when they force themselves on its attention.

I am not speaking of the experiments. What science says of the experiments is generally true. The experiments are very useful. It is the theorising about them which is so bad. Unfortunately, the practical instructions which science claims to give for various branches of life generally come from the theorising. You see how difficult it is to come to any understanding with this science, and yet—sooner or later we must do so. This understanding must be found, precisely for the most practical domains of life—and notably for Agriculture.

For all the different spheres of farming life we must gain insight into the working of the substances and forces, and of the Spiritual too. Such insight is necessary, so as to treat things in the right way. After all, a baby—so long as it does not know what a comb is for—will merely bite into it, treating it in an impossible and style-less fashion. We too shall treat things in an impossible and style-less fashion, so long as we do not know what their true essence is . . .

Consider a tree for example. A tree is different from an ordinary annual, which remains at the merely herbaceous stage. A tree surrounds itself with rind and bark, etc. What is the essence of the tree, by contrast to the annual? Let us compare such a tree with a little mound of earth which has been cast up, and which—we will assume—is very rich in humus, containing an unusual amount of vegetable matter more or less in process of decomposition, and perhaps of animal decomposition-products too. (Diagram 7).

Let us assume: this is the hillock of earth, rich in humus. And I will now make a hollow in it, like a crater. And let this (in the second drawing) be the tree: outside, the more or less solid parts, while *inside* is growing what leads eventually to the formation of the tree as a whole. It may seem strange to you that I put these two things side by side. But they are more nearly related than you would think.

In effect, earthly matter—permeated, as I have now described it, by humus-substances in process of decomposition—such earthly matter contains etherically living substance. Now this is the important point: Earthly matter, which by its special constitution reveals the presence in it of etherically living substance, is always on the way to become plant-integument. It only does not go far enough in the process to become such plant-integument as is drawn up, for instance, into the rind or bark of a tree.

You may conceive it thus (although in Nature it does not go so far): Imagine this hillock of earth being formed, with a hollow in the middle—a mound of earth, with humus entering into it, working in the earthly soil with the characteristic properties which proceed from the ethereal and living element. It does not happen so in Nature, but instead of it, the "mound of earth"—transmuted into a higher form of evolution—is gathered up around the plant so as to enclose it.

In effect, whenever in any given locality you have a general level or *niveau*, separating what is above the earth from the interior, all that is raised *above this normal level* of the district will show a special *tendency to life*—a tendency to permeate itself with *ethereal vitality*. Hence you will find it easier to permeate ordinary inorganic mineral earth with fruitful humus-substance, or with any waste-product in process of decomposition—you will find it easier to do this efficiently if you erect mounds of earth, and permeate these with the said substance. For then the earthly material itself will tend to become inwardly alive—akin to the plant-nature. Now the same process takes place in the forming of the tree. The earth itself is "hollowed upward" to surround the plant with its ethereal and living properties. Why so?

I am telling you all this to awaken in you an idea of the really intimate kinship between that which is contained within the contours of the plant and that which constitutes the soil around it. It is simply untrue that the life ceases with the contours—with the outer periphery of the plant. The actual life is continued, especially from the roots of the plant, into the surrounding soil. For many plants there is absolutely no hard and fast line between the life within the plant and the life of the surrounding soil in which it is living.

We must be thoroughly permeated with this idea, above all if we would understand the nature of manured earth, or of earth treated in some similar way. To manure the earth is to make it alive, so that the plant may not be brought into a dead earth and find it difficult, out of its own vitality, to achieve all that is necessary up to the fruiting process. The plant will more easily achieve what is necessary for the fruiting process, if it is immersed from the outset in an element of life. Fundamentally, all plant-growth has this slightly parasitic quality. It grows like a parasite out of the living earth. And it must be so.

In many districts, we cannot reckon upon Nature herself letting fall into the earth enough organic residues, and decomposing them sufficiently, to permeate the earth with the requisite degree of life. We must come to the assistance of plant-growth by manuring the earth. We need to do so least of all in those districts where "black earth," as it is called, prevails. For in "black earth"—at any rate in certain districts—Nature herself sees to it that the soil is sufficiently alive.

Thus we need to understand what is the essential point. But we must understand something else as well. We must know how to gain a kind of personal relationship to all things that concern our farming work, and above all—though it may be a hard saying—a personal relationship to the manure, especially to the task of working with the manure. It may seem an unpleasant task, but without this personal relation it is impossible. Why so? You will see it at once if you can go into the question: What is the essence of any living thing? A living thing always has an outer and an inner side. The "inner" is inside some kind of skin, the "outer" is outside the skin.

Consider now the inner side. It not only has streams of forces going outward in the direction of these arrows (Diagram 8); the inner life of an organic entity also includes currents of forces going *inward* from the skin—currents of forces that are pressed back. Moreover, outside it the organic entity is surrounded by manifold streams of forces.

Now there is something that expresses quite exactly—yet in a kind of personal way—how the organic entity establishes the right relationship between its inner and its outer side. All that goes on by way of forces and activities within it, stimulating and maintaining life within the organism—all that is inside the contours of the skin—all this (I beg you once more to forgive the hard saying) must *smell* inwardly, nay we might even say it must inwardly stink.

Life itself essentially consists in this, that what would otherwise scatter its scent abroad is held together, so that the aromatic elements do not ray outward too strongly, but are retained within. Towards the outer world, the organism must live in this way: through the contours of its skin it must let out as little as possible of that which engenders the scent-kindling life within it. So we might say: an organic body is the healthier, the more it *smells inwardly* and the less outwardly. Towards the outer world, the organism—notably the plant-organism—is predestined not to *give off* smell, but on the contrary to *absorb* it.

Perceive the helpful effect of a fragrant aromatic meadow, full of plants with aromatic scent! Then you become aware of the marvellous mutual aid prevailing in all life. The aromatic property which here expands and which is different from the mere aroma of life—it spreads its scent abroad for reasons which we may yet be able to describe, and it is this which works from without upon the plants.

These things we must have in a living and personal relationship; only then are we really *in* the life of Nature. The point is now to recognise the following. *Manuring* and everything of the kind consists essentially in this, that a certain degree of *livingness* must be communicated to the soil, and yet not only livingness. For the possibility must also be given to bring about in the soil what I indicated yesterday, namely to enable the *nitrogen* to spread out in the soil in such a way that with its help the life is carried along certain lines of forces, as I showed you. That is to say: in manuring we must bring to the earth-kingdom enough nitrogen to carry the living property to those structures in the earth-kingdom to which it must be carried—under the plant, where the plant-soil has to be. This is our task, and we must fulfil it in a scientific way.

There is one fact which can already give you a strong indication of what is needed. If you use mineral, purely mineral substances as manure, you will never get at the real earthy element; you will penetrate at most to the watery element of the earth. With *mineral manures* you can influence the *watery* content of the earth, but you do not penetrate sufficiently to bring to life the earth-element itself. Plants, therefore, which stand under the influence of mineral manures will have a kind of growth which betrays the fact that it is supported only by a quickened watery substance, not by a quickened earthy substance.

We can best approach these matters by turning, to begin with, to the most unassuming kind of manure. I mean the *compost*, which is sometimes even despised. In compost we have a means of kindling the life within the earth itself. We include in compost any kind of refuse to which little value is attached; refuse of farm and garden, from grass that we have let decay, to that which comes from fallen leaves or the like, nay, even from dead animals . . . These things should not by any means be despised, for they preserve something not only of the *ethereal* but even of the *astral*. And that is most important. From all that has been added to it, the compost-heap really contains ethereal and living elements and also astral. Living ethereal and astral elements are contained in it—though not so intensely as in manure or in liquid manure, yet in a more stable form. The ethereal and astral settle down more firmly in the compost; especially the astral.

The point is now to make use of this property in the right way. The influence of the astral on the nitrogen is marred in the presence of an all-too thriving ethereal element. Hypertrophy of the ethereal in the heap of compost does not give the astral a chance, so to speak. Now there is something in Nature, the excellence of which for Nature herself I have already described to you from several standpoints, and that is the *chalky* or *limestone* element. Bring some of this—

perhaps in the form of *quicklime*—into the heap of compost, and you will get this result: Without inducing the evaporation of the astral over-strongly, the ethereal is absorbed by the quicklime, and therewith oxygen too is drawn in, and the astral is made splendidly effective.

You thereby obtain quite a definite result. When you manure the soil with this compost, you communicate to it something which tends very strongly to permeate the earthy element with the astral, without going by the roundabout way of the ethereal. Think, therefore: the astral, without first passing via the ethereal, penetrates strongly into the earthy element. Thereby the earthy element is strongly astralised, if I may put it so, and through this astralising process is permeated by the nitrogen-content, in such a way that something arises very similar to a certain process in the human organism.

The process in the human organism to which I now refer is plant-like; plant-like, however, in the sense that it does not care to go on as far as the fruiting process, but is content to stop, as it were, at the stage of leaf- and stalk-formation. The process we here communicate to the Earth—we need it within us in order especially to bring into the foodstuffs that inner quickness and a mobility which, as I told you, is so necessary. And we shall kindle in the soil itself the same inner quickness and mobility if we treat it as I have now described. We then prepare the soil so that it brings forth something especially good for animals to consume; for in its further course it works in such a way that they develop inner mobility; their body becomes inwardly quick and alive.

In other words, we shall do well to manure our *meadows* and *pastures* with such compost. And if we do this properly—especially if we observe the other procedures which are necessary—we shall get very good pasture-food, good even as hay when it has been mown down. However, in order to proceed rightly in such matters we must always be able to see the whole. Our detailed measures must still depend on our inner feeling, to a large extent. This inner feeling will develop rightly, once we perceive the whole nature of the process.

For instance, if we just leave the pile of compost as I described it hitherto, it may easily come about that it will scatter its astral content on all sides. The point will be for us to develop the necessary personal relationship to these things. We must try to bring the compost-heap into such a condition that it smells as little as possible. This we can easily attain, to begin with, by piling it up in thin layers, covering it layer by layer with something else, for instance granulated peat, and then another layer and so on. That which would otherwise evaporate and scatter its scent abroad, is thereby held together. The nitrogen, in fact, is that which strongly tends to seek the wide expanse—in manifold forms and compounds. Now it is held together.

What I chiefly wish to indicate is that we must treat the whole agricultural life with the conviction that we need to pour vitality, nay even astrality, in all directions, so as to make it work as a totality.

Taking our start from this, another thing will result. Have you ever thought why cows have *horns*, or why certain animals have *antlers*? It is a most important question, and what ordinary science tells us of it, is as a rule one-sided and superficial. Let us then try to answer the question, why do cows have horns? I said just now that an organic or living entity need not only have streams of forces pouring outward: it can also have streams of forces pouring inward. Now imagine such an organic entity—of a lumpy and massive shape. It would have streams of forces going outward and streams of forces going inward. It would be very irregular; a lumpy organism—an ungainly creature. We should have strange-looking cows if this were all. They would be lumpy, with tiny appendages for feet, as indeed they are in the early embryonic stages. They would remain so; they would look quite grotesque.

But the cow is not like that. The cow has proper *horns* and *hoofs*. What happens at the places where the horns grow and the hoofs? A locality is formed which sends the currents *inward* with more than usual intensity. In this locality the outer is strongly shut off; there is no communication through a permeable skin or hair. The openings which otherwise allow the currents to pass outward are completely closed. For this reason the horn-formation is connected with the entire shaping of the animal. The forming of horns and hoofs is connected with the whole shape and form of the creature.

With the forming of *antlers* it is altogether different. Here the point is, not that the streams are carried back into the organism, but on the contrary, that certain streams are carried a certain way outward. There are valves, so to speak, whereby certain streams and currents are discharged outwardly. Such streams need not always be liquid or aeriform; they may also be currents of *forces*, localised in the antlers. The stag is beautiful because it has an intense communication with the surrounding world, inasmuch as it sends certain of its currents outward, and *lives* with its environment, thereby receiving all that works organically in the nerves and senses. So it becomes a quick and nervous animal. In a certain respect, all animals possessing antlers are filled with a gentle nervousness and quickness. We see it in their eyes.

The cow has horns in order to send into itself the astral-ethereal formative powers, which, pressing inward, are meant to penetrate right into the digestive organism. Precisely through the radiation that proceeds from horns and hoofs, much work arises in the digestive organism itself. Anyone who wishes to understand foot-and-mouth disease—that is, the reaction of the periphery on the digestive tract—must clearly perceive this relationship. Our remedy for foot-and-mouth disease is founded on this perception.

Thus in the horn you have something well adapted by its inherent nature, to ray back the living and astral properties into the inner life. In the horn you have something radiating life—nay, even radiating astrality. It is so indeed: if you could crawl about inside the living body of a cow—if you were there inside the belly of the cow—you would *smell* how the astral life and the living vitality pours inward from the horns. And so it is also with the hoofs.

This is an indication, pointing to such measures as we on our part may recommend for the purpose of still further enhancing the effectiveness of what is used as ordinary farm-yard-manure. What *is* farm-yard-manure? It is what entered as outer food into the animal, and was received and assimilated by the organism up to a certain point. It gave occasion for the development of *dynamic forces* and influences in the organism, but it was not primarily used to enrich the organism with material *substance*. On the contrary, it was excreted. Nevertheless, it has been inside the organism and has thus been permeated with an astral and ethereal content. In the astral it has been permeated with the nitrogen-carrying forces, and in the ethereal with oxygen-carrying forces. The mass that emerges as dung is permeated with all this.

Imagine now: we take this mass and give it over to the earth, in one form or another (we shall go into the details presently). What we are actually doing is to give the earth something ethereal and astral which has its existence by rights, inside the belly of the animal and there engenders forces of a plant-like nature. For the forces we engender in our digestive tract are of a *plant-like* nature. We ought to be very thankful that the dung remains over at all; for it carries astral and ethereal contents from the interior of the organs, out into the open. The astral and ethereal adheres to it. We only have to preserve it and use it in the proper way.

In the dung, therefore, we have before us something ethereal and astral. For this reason it has a life-giving and also astralising influence upon the soil, and, what is more, in the earth-element iself; not only in the watery; but notably in the earthy element. It has the force to overcome what is inorganic in the earthy element.

What we thus give over to the earth must of course have lost its original form, *i.e.*, the form it had before it was consumed as food. For it has passed through an organic process in the animal's digestive, metabolic system. In some sense it will be in process of dissolution and disintegration. But it is best of all if it is just at the point of dissolution by virtue of its own inherent ethereal and astral forces. Then come the little parasites—the minutest of living creatures—and find in it a good nutritive soil. These parasitic creatures are therefore generally supposed to have something to do with the goodness of the manure. In reality they are only *indicators* of the fact that the manure itself is in such and such a condition. As indicators of this they may well be of great importance; but we are under an illusion if we suppose that the manure can be fundamentally improved by inoculation with bacteria or the like. It may be so to outer appearance, but it is not so in reality. (I shall go into the matter at a later stage. Meanwhile, let us proceed).

We take manure, such as we have available. We stuff it into the horn of a cow, and bury the horn a certain depth into the earth—say about 18 in. to 2 ft. 6 in., provided the soil below is not too clayey or too sandy. (We can choose a good soil for the purpose. It should not be too sandy). You see, by burying the horn with its filling of manure, we preserve in the horn the forces it was accustomed to exert within the cow itself, namely the property of raying back whatever is life-giving and astral. Through the fact that it is outwardly surrounded by the earth, all the radiations that tend to etherealise and astralise are poured into the inner hollow of the horn. And the manure inside the horn is inwardly quickened with these forces, which thus gather up and attract from the surrounding earth all that is ethereal and life-giving.

And so, throughout the winter—in the season when the Earth is most alive—the entire content of the horn becomes inwardly alive. For the Earth is most inwardly alive in winter-time. All that is living is stored up in this manure. Thus in the content of the horn we get a highly concentrated, life-giving manuring force. Thereafter we can dig out the horn. We take out the manure it contains.

During our recent tests (in Dornach), as our friends discovered for themselves, when we took out the manure it no longer smelt at all. This was a very striking fact. It had no longer any smell, though naturally it began to smell a little when treated once more with water. This shows that all the odoriferous principles are concentrated and assimilated in it. Indeed it contains an immense ethereal and astral force; and of this you can now make use. When it has spent the winter in the earth, you take the stuff out of the horn and dilute it with ordinary water—only the water should perhaps be slightly warmed.

To give an impression of the quantitative aspect: I always found, having first looked at the area to be manured, that a surface, say, about as big as the patch from the third window here to the first foot-path, about 1,200 square metres (between a quarter- and third-acre) is adequately provided for if we use one hornful of this manure, diluted with about half a pailful of water. You must, however, thoroughly combine the entire content of the horn with the water. That is to say, you must set to work and stir. Stir quickly, at the very edge of the pail, so that a crater is formed reaching very nearly to the bottom of the pail, and the entire contents are rapidly rotating. Then quickly reverse the direction, so that it now seethes round in the opposite direction.

Do this for an hour and you will get a thorough penetration. Think, how little work it involves! The burden of work will really not be very great. Moreover, I can well image that—at any rate in the early stages—the otherwise idle members of a farming household will take pleasure in stirring the manure in this way. Get the sons and daughters of the house to do it and it will no doubt be wonderfully done.

It is a very pleasant feeling to discover how there arises after all, from what was altogether scentless to begin with, a rather delicately sustained aroma. This personal relationship to the matter (and you can well develop it) is extraordinarily beneficial—at any rate for one who likes to see Nature as a whole and not only as in the Baedeker guide-books.

Our next task will be to spray it over the tilled land so as to unite it with the earthly realm. For small surfaces you can do it with an ordinary syringe; it goes without saying, for larger surfaces you will have to devise special machines. But if you once resolve to combine your ordinary manuring with this kind of "spiritual manure," if I may call it so, you will soon see how great a fertility can result from such measures. Above all, you will see how well they lend themselves to further development. For the method I have just described can be followed up at once by another, namely the following.

Once more you take the horns of cows. This time, however, you fill them not with manure but with quartz or silica or even orthorclase or felspar, ground to a fine mealy powder, of which you make a mush, say of the consistency of a very thin dough. With this you fill the horn. And now, instead of letting it "hibernate," you let the horn spend the summer in the earth and in the late autumn dig it out and keep its contents till the following spring.

So you dig out what has been exposed to the summery life within the earth, and now you treat it in a similar way. Only in this case you need far smaller quantities. You can take a fragment the size of a pea, or maybe only the size of a pin's head, and distribute it by stirring it up well in a bucket of water. Here again, you will have to stir it for an hour, and you can now use it to sprinkle the plants externally. It will prove most beneficial with vegetables and the like.

I do not mean that you should water them with it in a crude way; you spray the plants with it, and you will presently see how well this supplements the influence which is coming from the other side, out of the earth itelf, by virtue of the cow-horn manure. And now, suppose you extend this treatment to the fields on a large scale. After all, there is no great difficulty in doing so. Why should it not be possible to make machines, able to extend over whole fields the slight sprinkling that is required? If you do this, you will soon see how the dung from the cow-horn drives from below upward, while the other draws from above—neither too feebly, nor too intensely. It will have a wonderful effect, notably in the case of cereals.

These things are derived from a larger sphere—not from what you do just at the moment with the single thing in hand, as though you would build up the entire human being theoretically from a single finger. No doubt, by such methods too, something is attained, which I by no means wish to under-estimate. Yet with all their investigations nowadays, people are trying to discover, as they put it, what is likely to be most productive for the farmer, and in the last resort it only amounts to this: they try to find how the production may be made financially most profitable. It really amounts to little more than that. The farmer may not always think of it; but unconsciously this is the underlying thought. He is astonished when by some measure he gets great results for the moment—say he gets big potatoes; or anything else that swells and has a comely size. But he does not pursue the investigation far enough beyond this point.

In effect, this is not at all the most important point. The important thing is, when these products get to man, that they should be beneficial for his life. You may cultivate some fruit of field or orchard in its appearance absolutely splendid, and yet, when it comes to man it may only fill his stomach without organically furthering his inner life. But the science of to-day is incapable of following the matter up to the point of finding how man shall get the best kind of nourishment for his own organism. It simply does not find the way to this.

How different it is in all that is here said out of Spiritual Science! Underlying it, as you have seen, is the entire household of Nature. It is always conceived out of the *whole*. Therefore each individual measure is truly applicable to the whole, and so it should be. If you pursue agriculture in this way, the result can be no other than to provide the very best for man and beast. Nay more, as everywhere in Spiritual Science, here too we take our start above all from *man* himself. Man is the foundation of all these researches, and the practical hints we give will all result from this. The end in view is the best possible sustenance of *human* nature. This form of study and research is very different from what is customary nowadays.

(The two Preparations mentioned in this lecture are now known as Preparations 500 and 501. The Preparations described in Lecture 5 are referred to in current literature as Preparations 502-507. During the past thirty-four years, the methods of making and applying the Preparations have been worked out, but quite intentionally, precise details have not been added to the present text because the Course of Lectures was intended to give *principles*, not technicalities, of their application. Further details of the method can be obtained by writing to the Bio-Dynamic Agricultural Association, Rudolf Steiner House, 35 Park Road, London, N.W.1.)

DISCUSSION

KOBERWITZ,
12th June, 1924.

Question: Should the dilution be continued arithmetically?

Answer: In this respect, no doubt, certain things will yet have to be discussed. Probably, with an increasing area you will need more water and proportionately fewer cow-horns. You will be able to manure large areas with comparatively few cow-horns. In Dornach we had twenty-five cow-horns; to begin with we had a fairly large garden to treat. First we took one horn to half a bucketful. Then we began again, taking a whole bucketful and two cow-horns. Afterwards we had to manure a relatively larger area. We took seven cow-horns and seven bucketfuls.

Question: Could one use a mechanical stirrer to stir up the manure for larger areas, or would this not be permissible?

Answer: This is a thing you can either take quite strictly, or else you can make up your mind to slide into substitute methods. There can be no doubt, stirring by hand has quite another significance than mechanical stirring. A mechanist, of course, will not admit it. But you should consider well what a great difference it makes, whether you really stir with your hand or in a mere mechanical fashion. When you stir manually, all the delicate movements of your hand will come into the stirring. Even the feelings you have may then come into it.

Of course the people of to-day will not believe that it makes any difference; but you can tell the difference even in medical matters. Believe me, it is not a matter of indifference whether a medicament is prepared more manually or mechanically. When a man works at a thing himself, he gives something to it which it retains. To mention one example, this is notably the case with the Ritter remedies, with which some of you are no doubt familiar. You must not smile at such things. I have often been asked what I think of the Ritter remedies. You are perhaps aware that there are some who sing hymns of praise on their behalf, while others spread the tale that they have no particular effect.

Undoubtedly they have an effect. But I am firmly convinced that if these remedies were brought on to the market in the usual way they would very largely lose their influence. With these remedies especially, it makes a great difference if the doctor himself possesses the remedy and gives it to his patient directly. When the doctor gives such a thing to his patient, when it is all taking place in a comparatively small circle, he brings a certain enthusiasm with him. You may say the enthusiasm as such weighs nothing; you cannot weigh it. Nevertheless it enters into the vibrations if the doctors are enthusiastic. Light has a strong effect on the remedies; why not enthusiasm? Enthusiasm mediates; it can have a great effect. Enthusiastic doctors of to-day can achieve great results. Precisely in this way, the Ritter remedies can have a far-reaching influence.

With enthusiasm, great effects can be called forth. But if you begin to do it in an indifferent and mechanical fashion, the effects will soon evaporate. It makes a difference whether you do the thing with all that proceeds from the human hand—believe me, very much can issue from the hand—or whether you do it with a machine. By and by, however, it might prove to be great fun—this stirring; and you would no longer dream of a mechanical stirrer even when many cow-horns were needed. Eventually, I can imagine, you will do it on Sundays as an after-dinner entertainment. Simply by having many guests invited and doing it on Sundays, you will get the best results without machines!

Question: No doubt there will be a little technical difficulty in distributing half a bucketful of water over one-fifth of an acre. But when you increase the number of cow-horns the difficulty will rapidly increase—quite out of proportion to the number. Can the given quantity of water be diluted still more, or is it essential to preserve the proportion of half a bucketful? Must you take about half a bucketful to one-fifth of an acre?

Answer: No doubt it will be possible as you suggest. But I think the method of stirring would then have to be changed. You might do it in this way. Stir up a cow-hornful completely in half a bucket of water, and then dilute it to a bucketful; but you will then have to stir it again.

On the whole, I think it would be best to stir only half a bucketful at a time. Reckon up, in the given instance, how much less of the stuff you need, even if it should be less than the contents of a cow-horn. It all depends on your bringing about a thoroughly intimate permeation. You are far from achieving a true permeation when you merely tip the stuff into water and stir it up a little. You must bring about a very intimate permeation. If you merely shake in the more or less condensed substance, or if you fail to stir it vigorously, you will not have a thorough mixture. Therefore I think it will be easier to stir several half-bucketfuls with small amounts of substance than to dilute the water again and stir it up a second time.

Question: Some solid matter will remain over, no doubt, even then. May the liquid afterwards be strained so that it can be distributed with a mechanical spray?

Answer: I do not think it will be necessary. For if you stir it quickly, you will obtain a fairly cloudy liquid, and you need not trouble whether any foreign bodies are left in it. You will not find it difficult to distribute the manure; pure cow-manure is best for the purpose, but even if there are foreign bodies in it, I do not think you need go to the trouble of cleansing it. If there are foreign bodies, they might even have a beneficial effect and do no harm. As a result of the concentration and subsequent dilution, it is only the radiant effect that works; it is no longer the substances as such, but the dynamic radiant activity. Thus there would be no danger, for example, of your getting potato plants with long shoots and nothing else upon them at the place where your foreign bodies happened to fall. I do not think there would be any such danger.

Question: I only had in mind the mechanical spray.

Answer: Certainly you can strain the liquid; it will do it no harm. It might be simplest to have your mechanical spray fitted with a sieve from the outset.

Question: You did not say whether the stuff from the horn should be weighed out, so as to get a definite proportion. Speaking of half a bucketful, did you refer to a Swiss bucket, or a precise measure of litres?

Answer: I took a Swiss bucket, the ordinary bucket they use for milking in Switzerland. The whole thing was tested practically, in the direct perception of it. You should now reduce it to the proper weights and measures.

Question: Can the cow-horns be used repeatedly, or must they always be taken from freshly slaughtered beasts?

Answer: We have not tested it, but from my general knowledge I think you should be able to use the cow-horns three or four times running. After that they will no longer work so well. There might even be this possibility: Use the cow-horns for three or four years in succession; then keep them in the cow-stable for a time, and use them again another year. This too might be possible. But I have no idea how many cow-horns an agricultural area can normally have at its disposal; whether or no it is necessary to be very economical in this respect. That is a question I cannot decide at the moment.

Question: Where can you get the cow-horns? Must they be taken from Eastern-European or Mid-European districts?

Answer: It makes no difference where you get them from—only not from the refuse yard. They must be as fresh as possible. However, strange as it may sound, it is a fact that Western life—life in the Western hemisphere—is quite a different thing from life in the Eastern hemisphere. Life in Africa, Asia or Europe has quite another significance than life in America. Possibly, therefore, horns from American cattle would have to be made effective in a rather different way. Thus it might prove necessary to tighten the manure rather more in these horns—to make it denser, hammer it more tightly.

It is best to take horns from your own district. There is an exceedingly strong kinship between the forces in the cow-horns of a certain district and the forces generally prevailing in that district. The forces of horns from abroad might come into conflict with what is there in the earth of your own country. You must also remember, it will frequently happen that the cows from which you get the horns in your own district are not really native to the district. But you can get over this difficulty. When the cows have been living and feeding on a particular soil for three or four years, they belong to the soil (unless they happen to be Western cattle).

Question: How old may the horns be? Should they be taken from an old or a young cow?

Answer: All these things must be tested. From the essence of the matter, I should imagine that cattle of medium age would be best.

Question: How big should they be?

Answer: Dr. Steiner draws on the board the actual size of the horn—about 12 to 16 inches long (Diagram 9), *i.e.* the normal size of horn of " Allgäu " cattle, for example.

Question: Is it not also essential whether the horn is taken from a castrated ox, or from a male or female animal?

Answer: In all probability the horn of the ox would be quite ineffective, and the horn of the bull comparatively weak. Therefore I speak of cow-horns; cows as a rule are female. I mean the female animal.

Question: What is the best time to plant cereals?

Answer: The exact answer will be given when I come to sowing in the main lectures. It is very important, needless to say, and it makes a great difference whether you do it more or less near to the winter months. If near to the winter months, you will bring about a strong reproductive power in your cereals; if farther from the winter months, a strong nutritive power.

Question: Could the cow-horn manure also be distributed with sand? Is rain of any importance in this connection?

Answer: As to the sand you may do so; we have not tested it, but there is nothing to be said against it. The effect of rain would also have to be tested. Presumably it would bring about no change; it might even tend to establish the thing more firmly. On the other hand, we are dealing with a very high concentration of forces, and possibly the minute impact of the falling raindrops might scatter the effect too much. It is a very delicate process; everything must be taken into account. There is nothing to be said against spreading sand with the cow-manure.

Question: In storing the cow-horns and their contents, how should one prevent any harmful influences from gaining access?

Answer: In these matters it is generally true to say that you do more harm by removing the harmful influences, so-called, than by leaving them alone. Nowadays, as you know, people are always wanting to " disinfect " things. Undoubtedly they go too far in this. With our medicaments, for example, we found that if we wished absolutely to prevent the possibility of mould, we had to use methods which interfere with the real virtue of the medicament.

I for my part have no great respect for these " harmful influences." They do not do nearly so much harm. The best thing is, not to go out of our way in devising methods of purification, but to let well alone.

(We only put pig's bladder over the top to prevent the soil from falling in.)

To try to clean the horns by any special methods is not at all to be recommended. We must familiarise ourselves with the fact that " dirt " is not always dirt. If, for example, you cover your face with a thin layer of gold, it is " dirt " ; and yet, gold is not dirt. Dirt is not always dirt. Somethimes it is the very thing that acts as a preservative.

Question: Should the extreme " chaoticizing " of the seed, of which you spoke, be supported or enhanced by any special methods?

Answer: You could do so, but it would be superfluous. If the seed-forming process occurs at all, the maximum of chaos will come of its own accord. There is no need to support it. It is in manuring that the support is needed. In the seed-forming process, I do not think it will be necessary to enhance the chaos any more. If there is fertilising seed at all, the chaos is complete. You could do it, of course, by making the soil more silicious. It is through silica that the essential cosmic forces work.

Whatever cosmic forces are caught up by the earth, work through the silica. You could do it in this way, but I do not believe it is necessary.

Question: How large should the experimental plots be? Will it not also be necessary to do something for the cosmic forces that should be preserved until the new plant is formed?

Answer: You might experiment as follows. It is comparatively easy to give general guiding lines; but the most suitable scale on which to work is a thing you must test for yourselves. It will not, however, be difficult to make experiments on this question. Set out your plants in two separate beds, side by side—a bed of wheat, say, and a bed of sainfoin. Then you will find this possibility. In the one plant—wheat—which of its own accord tends easily to lasting seed-formation, you will retard the seed-forming process by the use of silica. Meanwhile, with the sainfoin, you will find the seed-forming process quite suppressed or very much retarded.

To investigate these things, you can always take this as a basis of comparison: Study the properties of cereals—wheat, for example—and then compare them with the analogous properties of sainfoin, or leguminosae generally. You will thus have the most interesting experiments on seed-formation.

Question: Does it matter *when* the diluted stuff is brought on to the fields?

Answer: Undoubtedly it does. You can generally leave the cow-horns in the earth until you need them. They will not deteriorate, even if after hibernating they are left for a while during the summer. If, however, you do need to keep them elsewhere, having taken them out of the earth, you should make a box, upholster it well with a cushion of peat-moss on all sides, and put the cow-horns inside. Then the strong inner concentration will be preserved. In any case, it is inadvisable to keep the watery fluid after dilution. You must do the stirring not too long before you use the liquid.

Question: If we want to treat the winter corn, must we use the cow-horns a whole quarter after taking them out of the earth?

Answer: It does not matter essentially, but it will always be better to leave them in the earth until you need them. If you are going to use them in the early autumn, leave them in the earth until you need them. It will in no way harm the manure.

Question: With the fine spraying of the liquid due to the spraying machine, will not the etheric and astral forces be wasted?

Answer: Certainly not; they are intensely bound. Altogether, when you are dealing with spiritual things—unless you drive them away yourself from the outset—you need not fear that they will run away from you nearly as much as with material things.

Question: How should one treat the cow-horns with mineral content, after they have spent the summer in the earth?

Answer: It will not hurt to take them out and keep them anywhere you like; you can throw them in a heap anywhere. It will not hurt the stuff, when it has once spent the summer in the earth. Let the sun shine on them; it will not hurt, it will even do them good.

Question: Must the horns be buried at the same place—on the same field which you will afterwards be wanting to manure, or can they be buried all together at any place you choose?

Answer: It makes so little difference that you need not worry about it. In practice, it will be best to look for a place where the soil is comparatively good. I mean, where the earth is not too highly mineral, but contains plenty of humus. Then you can bury all the cow-horns you need in one place.

Question: What about using machines on the farm? Is it not said that machines should not be used at all?

Answer: That cannot really be answered purely as a farming question. Within the social life of to-day, it is hardly a practical, hardly a topical question to ask whether machines are allowable. You can hardly be a farmer nowadays without using machines. Needless to say, not all operations are so nearly akin to the most intimate processes of Nature as the stirring of which we were speaking just now. Just as we did not want to mix up such an intimate process of Nature with purely mechanical elements, so it is with regard to the other things of which you are thinking. Nature herself, in any case, sees to it that where machines are out of place you can do very little with them. A machine will not help in the seed-forming process, for example; Nature does it for herself.

Really I think the question is not very practical. How can you do without machines nowadays? On the other hand, I may remark that as a farmer you need not just be crazy on machines. If one has a particular craze for machines, he will undoubtedly do worse as a farmer, even if his new machine is an improvement, than if he goes on using his old machine until it is worn out. However, in the strict sense of the word these are no longer purely farming questions.

Question: Could the given quantity of cow-horn manure, diluted with water, be used on half the area you indicated?

Answer: Then you would get rampant growths; you would get the result I hinted at just now in another connection. If, for example, you did this in potato-growing or the like, you would get rampant plants with highly ramified stems; what you are really wanting would not develop properly. Apply the stuff in excess and you will get what are generally known as rank patches.

Question; What about a fodder plant, which you want to grow rampant—spinach for instance?

Answer: There, too, I think we shall only use the half-bucketful with the one cow-horn. That is what we did in Dornach with a patch that was mainly vegetable garden. For plants that are grown over larger areas, you will need far less in proportion. It is already the optimum amount.

Question: Does it matter what kind of manure you use—cow- or horse- or sheep-manure?

Answer: Undoubtedly cow-manure is best for this procedure. Still, it might also be well to investigate whether or no horse-manure could be used. If you want to treat horse-manure in this way, you will probably find that you need to wrap the horn up to some extent in horse-hair taken from the horse's mane. You will thus make effective the forces which in the horse—as it has no horns—are situated in the mane.

Question: Should it be done before or after sowing the seed?

Answer: The proper thing is to do it before. We shall see how it works; this year we began rather late, and some things will be done after sowing. We shall see whether it makes any difference. However, as a normal matter of course, you should do it before sowing, so as to influence the soil itself beforehand.

Question: Can the same cow-horns that have been used for manure be used for the mineral substance too?

Answer: Yes, but here too you cannot use them more than three or four times. After that they lose their forces.

Question: Does it matter who does the work? Can anyone you choose do the work, or should it be an anthroposophist?

Answer: That is the question. If you raise such a question at all nowadays, you will be laughed at, no doubt, by many people. Yet I need only remind you that there are people whose flowers, grown in the window-box, thrive wonderfully, while with others they do not thrive at all but fade and wither. These are simple facts.

These things that take place through human influence, though they cannot be outwardly explained, are inwardly quite clear and transparent. Moreover, such things will come about simply as a result of the human being practising meditation; preparing himself by meditative life, as I described it in yesterday's lecture. For when you meditate you live quite differently with the nitrogen which contains the Imaginations. You thereby put yourself in a position which will enable all these things to be effective; you put yourself in this position over against the whole world of plant-growth.

However, these things are no longer as clear to-day as they used to be in olden times, when they were universally accepted. For there were times when people knew that by certain definite practices they could make themselves fitted to tend the growth of plants. Nowadays, when such things are not observed, the presence of other people disturbs them. These delicate and subtle influences are lost when you are constantly living and moving among men and women who take no notice of such things. Hence, if you try to apply them, it is very easy to prove them fallacious. And I am loth to speak openly as yet about these things in a large company of people. The conditions of life nowadays are such that it is only too easy to refute them.

A very ticklish question was raised, for example, by our friend Stegemann in the discussion in the Hall the other day, namely, whether parasites could be combatted by such means—by means of concentration or the like. There can be no question about it that you can, provided you did it in the right way. Notably you would want to choose the proper season—from the middle of January to the middle of February—when the earth unfolds the greatest forces, the forces that are most concentrated in the earth itself. Establish a kind of festival time, and practise certain concentrations during the season, and the effects might well be evident.

As I said, it is a ticklish question, but it can be answered positively along these lines. The only condition is that it must be done in harmony with Nature as a whole. You should be well aware that it makes all the difference whether you do an exercise of concentration in the winter-time or at midsummer. How much is contained in many of the old folk-proverbs! Even the people of to-day might still derive many a valuable hint from these.

I could have mentioned it in yesterday's lecture: Among the many things I should have done in this present incarnation, but did not find it possible to do, was this. When I was a young man I had the idea to write a kind of " peasant's philosophy," setting down the conceptual life of the peasants in all the things that touch their lives. It might have been very beautiful. The statement of the Count, that peasants are stupid, would have been refuted. A subtle wisdom would have emerged—a philosophy dilating upon the intimacies of Nature's life—a philosophy contained in the very formation of the words. One marvels to see how much the peasant knows of what is going on in Nature.

To-day, however, it would no longer be possible to write a peasant's philosophy. These things have been almost entirely lost. It is no longer as it was fifty or forty years ago. Yet it was wonderfully significant; you could learn far more from the peasants than in the University. That was an altogether different time. You lived with the peasants in the country, and when those people came along with their broad-brimmed hats, introducing the Socialist Movement of to-day, they were only the eccentricities of life. To-day the whole world is changed. The younger ladies and gentlemen here present have no idea how the world has changed in the last thirty or forty years. How much has been lost of the true peasants' philosophy, of the real beauty of the folk-dialects! It was a kind of cultural philosophy.

Even the peasants' calendars contained what they no longer contain to-day. Moreover, they looked quite different—there was something homely about them. I, in my time, knew peasants' calendars printed on very poor paper, it is true ; inside, however, the planetary signs were painted in colours, while on the cover, as the first thing to meet the eye, there was a tiny sweet which you might lick whenever you use the book. In this way too it was made tasty; and of course the people used it one after another.

Question: When larger areas are to be manured, must the number of cow-horns be determined purely by feeling?

Answer: No, I should not advise it. In such a case, I think, we really must be sensible. This, therefore, is my advice. Begin by testing it thoroughly according to your feeling. When you have done all you can to get the most favourable results in this way, then set to work and translate your results into figures for the sake of the world as it is to-day. So you will get the proper tables which others can use after you.

If anyone is inclined to do it out of pure feeling, by all means let him do so. But in his attitude to others he should not behave as though he did not value the tables. The whole thing should be translated into calculable figures and amounts for the sake of others; it is necessary nowadays. You need cows' horns to do it with, but you do not exactly need to grow bulls' horns in representing it! These are the things that lead so easily to opposition. I should advise you as far as possible to compromise in this respect, and bear in mind the judgments of the world at large.

Question: Is the quick-lime treatment of the compost-heap, in the percentages as given nowadays, to be recommended?

Answer: The old method will undoubtedly prove beneficial, only you must treat it specifically, according to the nature of your soil —whether it be more sandy or marshy. For a sandy soil you will need rather less quicklime. A marshy ground will need rather more quicklime on account of the formation of oxygen.

Question: How about digging up and turning over the compost-heap?

Answer: That is not bad for it. When you have dug it up and turned it, you should, however, provide for its proper protection by putting a layer of earth all around it. Cover it over with earth; peat-earth or granulated peat is very good for the purpose.

Question: What kind of potash did you mean, when you said it *might* be used if necessary in the transition stage?

Answer: Kali magnesia.

Question: What is the best way of using the rest of the manure after the cow-horns have been filled? Should it be brought on to the fields in autumn, so as to undergo the winter experience? or should it be set aside until the spring?

Answer You must remember that the cow-horn manuring is not intended as a complete substitute for ordinary manuring. You should go on manuring as before. The new method should be regarded as a kind of extra, largely enhancing the effect of the manuring hitherto applied. The latter should continue as before.

LECTURE FIVE

KOBERWITZ,
13th June, 1924.

MY DEAR FRIENDS,

The preparation I indicated yesterday for the improvement of manure was intended, of course, simply as an improvement, as an enhancement. Needless to say, you will go on manuring as before. To-day we shall have to consider the manuring problem still further, in view of our necessary standpoint that whatever is living must be kept within the living sphere. Ethereal life, as we have seen, should never depart from anything that is in the sphere of living growth. Hence it was of great value for us to recognise that the soil out of which the plant grows and which surrounds the roots, is in itself a kind of continuation of growth within the earth. There is a vegetative plant-life in the earth itself.

In yesterday's lecture I even shewed how we can imagine the transition from a thrown-up hillock of earth—with the inner vitality of its humus-content—to the rind or even the bark that surrounds the tree, enclosing the tree from the outside. Naturally enough, in modern time, when all insight into the great connections of Nature has been lost—as indeed it had to be—this insight too has gone. Science no longer perceives this common life—common to the Earth and all plant-growth—nor how it is continued into the excretion-products of life in the manure. Science no longer knows the working of this all-embracing life. Insight into these things *had* to be lost, increasingly as time went on.

Now Spiritual Science, as I said in yesterday's discussion, must not come in in a turbulent and revolutionary spirit, interfering with all that our time has achieved in the different domains of life. We must begin by recognising what has really been achieved. We must oppose or fight those things alone which rest on completely false premises—which are a mere outcome of the materialistic world-conception. Meanwhile, in all the different spheres of life, we must try to supplement genuine modern achievement with that which can flow from our own, living conception of the Universe.

Therefore I need not spend much time describing how you should prepare manure—whether from stable manure, liquid manure or compost. In this respect—for the due preparation of manure and liquid manure—much has already been done. Perhaps we can say more of these things in this afternoon's discussion. I will only say this to begin with: The idea that in farming we are really *exploiting* the land is quite correct. Indeed, we cannot help doing so. With all that we send out into the world from our farms, we are taking forces away from the earth—nay, even from the air. These forces must somehow be restored. After a time, the manure-substance whose inner value is so deeply connected with all that we need for the impoverished earth, must be subjected to a proper treatment, so as to quicken and vitalise it sufficiently.

Notably in the most recent times, many false judgments have arisen from the materialistic outlook in this respect. They are at pains to investigate the working of bacteria—the smallest of living entities. They ascribe to these minute creatures the virtue of preparing the right conditions and relationships of substance in the manure. They reckon first and foremost on all that the bacteria do for the manure. Brilliant, highly logical experiments have been made, inoculating the soil with bacteria. Truly brilliant! but as a rule they have not stood the test of time, for they have proved of little use.

These things, in fact, are done from a point of view for which the following is a just parallel: Here is a room; we find an extraordinary number of flies in it. Because there are so many flies, we say the room is dirty. But the room is not dirty because of the flies. On the contrary, the flies are there because the room is dirty. Nor should we clean the room by thinking out devices to increase the number of flies (imagining that they will eat the dirt up more quickly) or even to diminish them, or anything of that kind. We shall attain far more by tackling the dirt itself, directly.

So it is when we use animal excretion-products as manure. We must regard the minute living entities as occurring by virtue of the processes that arise of themselves, here or there in the dung-substance. The presence of these creatures may therefore be an extremely useful symptom of the prevalence of such and such conditions in the dung-substance itself. But there can be no great good in planting them or breeding them. (Indeed, we might often do more good by combatting them). In effect, for the living life which is so vital to agriculture, we should always remain in *larger* spheres, and even to these minutest of creatures we should apply as little as possible of atomistic forms of thought.

It should go without saying that such a statement ought never to be made unless we are able to shew positive ways and means at the same time. No doubt, what I have now been saying is emphasized in many quarters. But it is not only important to know what is abstractly correct. If our correct knowledge is merely negative it generally helps us little; we must have positive principles to set over against it. That is the point in every case! If positive proposals cannot be made, we had better refrain from stressing the negative, for it will only tend to annoy.

A second thing is this: As a result of materialistic tendencies, once more it has been thought well in modern times to treat the manure in various ways with inorganic substances—compounds or elements. Here too, however, people are learning from experience. It has no permanent value. We must in fact be clear on this: So long as we try to ennoble or improve the manure by mineralising methods, we shall only succeed in quickening the liquid element—the water. Now for a firm and sound plant-structure it is necessary not only to quicken and organise the water—for from the water which merely trickles through the earth, no further vitalisation proceeds.

We must vitalise the *earth* directly, and this we cannot do by merely mineral procedures. This we can only do by working with *organic* matter, bringing it into such a condition that it is able to organise and vitalise the solid earthy element itself. To endow the mass of manure, or the liquid manure, with this kind of quickening or stimulus, is precisely the object of those inspirations which we are able to give to agriculture out of spiritual science. This quickening, this stimulation, can be given to any mass that is available as manure, provided always we remain within the sphere of life.

Spiritual Science always tries to look into the effects of living things on a *large* scale. It does not pry into the minute and microscopic, for that is not the most important. It does not primarily concern itself with the conclusions which are drawn from the minute—from microscopical investigations. To observe the *macrocosmic*—the wide circumference of Nature's workings—that is the task of Spiritual Science. But we must first know how to penetrate into these wider workings of Nature.

There is a saying you will often find repeated in agricultural literature, in many variations. No doubt it arises from the experiences which they believe they have collected. It is to this effect: "Nitrogen, phosphoric acid, calcium, potash, chlorine, etc., even iron—all these are essential in the soil if plant-growth is to prosper there. Silicic acid, on the other hand, lead, arsenic, mercury"—and they even include soda in this category—"have for plant-life at most the value of stimulants or irritants. One may stimulate the plants with them, but that is all." In this very statement, the men of to-day betray the fact that they are really groping about in the dark. It is a very good thing—as a result of tradition, no doubt—that they do not treat the plants as madly as they would do if they really followed this proposition. It is, as a matter of fact, impossible to do so.

What is the truth in this connection? Great Nature does not leave us so mercilessly in the lurch if we fail to take the silicic acid or the lead or mercury or arsenic into account, as she does if we fail to take into account her potash or limestone or phosphoric acid. Heaven provides silicic acid, lead, mercury, and arsenic—provides them freely with the rain. On the other hand, to have the proper phosphoric acid, potash and limestone-content in the Earth, we must till the soil and manure it properly. Heaven does not give these things of her own accord.

Nevertheless, by prolonged tillage we *can* gradually impoverish the soil. We are, of course, constantly impoverishing it, and that is why we have to manure it. But the compensation through the manure may presently become inadequate—and this is happening to-day on many farms. Then we are ruthlessly exploiting the earth; we let it become permanently impoverished. We must then provide for the true Nature-process to take place once more in the right way.

Those that are commonly called the *stimulant* effects are indeed the *most important* of all. Precisely the substances people think inessential are present all around the Earth—actively working, though in the finest and most tenuous dilution. Moreover, the plants need them just as much as they need what comes to them from the Earth. They draw them in from the world-circumference—from the cosmic circle. Mercury, arsenic, silicic acid—these substances the plants suck upward from the soil of the Earth after they have been rayed into the soil from the Cosmos.

However, we as human beings can utterly prevent the soil's receiving from the world-circumference, and raying outward in the proper way, what the plants need in this respect. If we continue manuring at random from year to year, we *can* gradually prevent the Earth from drawing into itself what it needs by way of silicic acid, lead and mercury, which are at work in the finest homoeopathic doses, if I may put it so—coming inward from the world-circumference. These influences need to be absorbed into the growth of the plant, if it is really to receive all that it needs from the Earth. For with the help of all that comes from the world-circumference in this fine and delicate condition, the plant builds up its body in the configuration of carbon.

Therefore we need to treat our manure not only as I indicated yesterday; we should also subject it to a further treatment. And the point is not merely to add substances to it, with the idea that it needs such and such *substances* so as to give them to the plants. No, the point is that we should add *living forces* to it. The living forces are far more important for the plant than the mere substance-forces or substances. Though we might gradually get our soil ever so rich in this or that substance, it would still be of no use for plant-growth, unless by a proper manuring process we endowed the plant itself with the power to receive into its body the influences which the soil contains. This is the point.

The men of our time are altogether unaware how the minutest quantities will often work with great intensity, precisely where living things are concerned. Now, however, we have the brilliant investigations of Frau Dr. Kolisko on the effects of "smallest entities." What hitherto, in homoeopathy, was a blind groping in the dark, has here been placed on a sound scientific footing, and as an outcome of her work I think we may take it as proved that in the minute entities, in the minute quantities, the radiant forces we need in the organic world *are* really set free—provided only that we use these entities in the proper way. And in manuring it is not at all difficult for us to use the minute quantities in the proper way.

You will remember how we prepare the forces in the cow's horns, and how we add the preparations, as the case may be, before or after manuring. These forces and influences then assist the working of the manure itself. We add these forces, so as to assist the working of the manure, which, apart from these homoeopathic doses, is used in the proper way, as heretofore. But in other ways, too, we must still try to give the manure the right living property. We must give it such a consistency that it will retain of its own accord as much of nitrogen and other substances as it requires. For we shall thereby impart to the manure a tendency to that living vitality which will enable it to bring the right vitality into the Earth itself.

To-day therefore—more as a general indication—I shall mention a few more things in the same direction: preparations to add to the manure in minute doses, in addition to the cow-horn stuff. The preparations we add to the manure vitalise it in such a way that it will then be able to transmit its vitality to the soil from which the plants are springing.

I shall mention various things, but let me say at the outset: if they should be difficult to obtain in one district or another, they can, if need be, be replaced by certain other things. Only in one case a substitute cannot be found, for it is so characteristic that the effect is scarcely likely to be found in the same way in any other plant.

From what I have said hitherto, we must provide for those things of the Universe which are above all important—namely, carbon, hydrogen, nitrogen and sulphur—to come together in the right way with other substances in the organic realm; notably with *potash* salts, for instance. As to the mere quantity of potash salts which the plant needs for its growth, no doubt a little of these things is already known. It is well-known that potash-salts (or potash, generally speaking) carry the growth rather into those regions of the plant organism which become rigid structure or framework in many instances, *i.e.* which bring about the formation of trunk or stem or the like. The potash-content will hold back the growth in forming strong and sturdy stems, etc. But it is very important—in all that takes place as between the earth and the plant—so to assimilate the potash-content that it relates itself rightly, within the organic process, to that which really constitutes the body of the plant, *i.e.* to the protein substance. Here we shall be successful if we proceed as follows:—

Take *yarrow**—a plant which is generally obtainable. If there is none of it in the district, you can use the dried herb just as well. Yarrow is indeed a miraculous creation. No doubt every plant is so; but if you afterwards look at any other plant, you will take it to heart all the more, what a marvel this yarrow is. It contains that of which I told you that the Spirit always moistens its fingers therewith when it wants to carry the different constituents—as carbon, nitrogen, etc.—to their several organic places. Yarrow stands out in Nature as though some creator of the plant-world had had it before him as a model, to shew him how to bring the *sulphur* into a right relation to the remaining substances of the plant.

One would fain say, " In no other plant do the Nature-spirits

* *Achillea millefolium*—also known as Milfoil.

attain such perfection in the use of sulphur as they do in yarrow." And if you also know of the working of yarrow in the animal or human organism—if you know how well it can make good all that is due to weaknesses of the astral body (provided it is rightly carried into the biological sphere)—then you will trace it still farther, in its yarrow-nature, throughout the entire process of plant growth. Yarrow is always the greatest boon, wherever it grows wild in the country—at the edges of the fields or roads, where cereals or potatoes or any other crops are growing. It should on no account be weeded out. (Needless to say, we should prevent it from settling where it becomes a nuisance—it may become a nuisance, though it is never actually harmful).

In a word, like sympathetic people in human society, who have a favourable influence by their mere presence and not by anything they say, so yarrow, in a district where it is plentiful, works beneficially by its mere presence.

Now you can do the following. Take the same part of the yarrow which is medicinally used, namely, the upper part—the umbrella-shaped inflorescence. If you have yarrow ready to hand, so much the better. Pick the fresh flowers and let them dry, only for a short time. Indeed, you need not let them dry so very much. If fresh yarrow is unobtainable—if you can only get the dried herb—you will do well before using it to press the juice out of the yarrow leaves. (Even from the dried leaves, you can get the required juice by decoction). Water the inflorescence a little with this juice.

Now you will see once more how we always remain within the living sphere. Take one or two hollow handfuls of this yarrow-stuff, pressed pretty strongly together, and sew it up in the bladder of a stag. Enclose the yarrow substance as best you can in the stag's bladder, and bind it up again. There, then, you have a fairly compact mass of yarrow in the stag's bladder. Now hang it up throughout the summer in a place exposed as far as possible to the sunshine. When autumn comes, take it down again and bury it not very deep in the Earth throughout the winter.

So you will have the yarrow flower (it matters not if it be tending already towards the fruit) enclosed in the bladder of the stag for a whole year, and exposed—partly above the earth, partly below—to those influences to which it is susceptible. You will find that it assumes a peculiar consistency during the winter.

In this form you can now keep it as long as you wish. Add the substance which you take out of the bladder to a pile of manure—it may even be as big as a house!—and distribute it well. Nay, you need not even do much to distribute it: the radiation itself will do the work. The radiating power is so very strong that if you merely put it in —even if you do not distribute it much—it will influence the whole mass of manure or liquid manure or compost. (If we speak of radiating forces, the materialists will believe us, will they not, for even they speak of radium!)

The mass we thus gain from the yarrow has an effect so quickening and so refreshing that if we now use the manure thus treated, just in the way manure is ordinarily used, we shall make good again much that would otherwise become a ruthless exploitation of the earth. We re-endow the manure with the power, so to quicken the earth that the more distant *cosmic* substances—silicic acid, lead, etc., which come to the earth in finest homoeopathic quantities—are caught up and received.

Here again the members of our Agricultural Circle should make experiments; they will soon see how well it works. And now the question is (for we should always work with insight, not with lack of insight), the question is: As to the yarrow, we have learned to know it. Its homoeopathic sulphur-content, combined in a truly model way with potash, not only works magnificently in the plant itself, but enables the yarrow to ray out its influences to a greater distance and through large masses. But the question remains: Why should we sew it up precisely in the bladder of a stag?

Here we must gain an insight into the whole process that is connected with the *bladder*. The stag is an animal most intimately related, not so much to the Earth but to the Earth's environment, *i.e.* to the Cosmic in the Earth's environment. Therefore the stag has antlers, the functions of which I explained yesterday. Now that which is present in the yarrow is intensely preserved, both in the human and in the animal organism, by the process which takes place between the kidneys and the bladder. Moreover, this process itself is dependent on the substantial nature or consistency of the bladder. Thus, in the bladder of the stag—however thin it is in substance—we have the necessary *forces*. Unlike the former instance (the cow, which is quite different), these forces are not connected with the interior. The bladder of the stag is connected rather with the forces of the Cosmos. Nay, it is almost an image of the Cosmos. We thereby give the yarrow the power quite essentially to enhance the forces it already possesses, to combine the sulphur with the other substances.

In this yarrow treatment we have an absolutely fundamental method of improving the manure, while all the time we remain within the realm of living things. We never go out of the living realm into that of inorganic chemistry. This is important to observe.

Now take another example. We want to give the manure the power to receive so much life into itself that it is able to transmit life to the soil out of which the plant is growing. But we must also make the manure able to bind together, still more, the substances which are necessary for plant growth—that is, in addition to potash, also the *calcium* compounds. In yarrow we are mainly dealing with potassium influences. If we also wish to get hold of the calcium influences, we need another plant, which—if it does not enthuse us like yarrow—also contains sulphur in homoeopathic quantity and distribution, so as to attract through the sulphur the other substances which the plant needs, and draw them into an organic process.

This plant is *camomile* (*Chamomilla officinalis*). It is not enough to say that camomile is distinguished by its strong potash and calcium contents. The facts are these: Yarrow mainly develops its sulphur-force in the potash-formative process. Hence it has sulphur in the precise proportions which are necessary to assimilate the potash. Camomile, however, assimilates calcium in addition. Therewith, it assimilates that which can chiefly help to exclude from the plant those harmful effects of fructification, thus keeping the plant in a healthy condition. It is a wonderful thing to see. Camomile too has a certain amount of sulphur in it, but in a different quantity, because it has calcium to assimilate as well.

Now once again you can look around you. The indications of Spiritual Science invariably consider the great and wide circles of life—the macrocosmic, not the microscopic conditions. Now you must trace, for example, the process which camomile undergoes in the human and animal organism, when taken as food or medicine. The bladder is comparatively unimportant for what the camomile must undergo in the human or animal organism. In this case, the substance of the intestinal walls is far more important. Therefore, if you want to work with camomile—as is the other case with yarrow—you must proceed as follows.

Pick the beautiful delicate little yellow-white heads of the flowers, and treat them as you treated the umbels of the yarrow. But now, instead of putting them in a bladder, stuff them into *bovine intestines*. You will not need very much. Here again, it is a charming operation. Instead of using these intestinal tubes as they are commonly used for making sausages, make them into another kind of sausage—fill them with the stuffing which you thus prepare from the camomile flower.

This preparation, once more, need only be rightly exposed to the influences of Nature. Observe how we constantly remain within the living realm. In this case, living vitality connected as nearly as possible with the *earthy* nature must be allowed to work upon the substance. Therefore you should take these precious little sausages —for they are truly precious—and expose them to the earth throughout the winter. Bury them not too deep, in soil as rich as possible in humus. If possible, choose a spot where the snow will remain for a long time and where the sun will shine upon the snow, for you will thus contrive to let the cosmic astral influences work down into the soil where your precious little sausages are buried.

Dig them out in the springtime and keep them in the same way as before. Add them to the manure just as you did the yarrow-preparation. You will thus get a manure with a more stable nitrogen-content, and with the added virtue of kindling the life in the earth, so that the earth itself will have a wonderfully stimulating effect on the plant-growth. Above all, you will create more healthy plants—really more healthy—if you manure in this way than if you do not.

I know perfectly well, all this may seem utterly mad. I only ask you to remember how many things have seemed utterly mad, which have none the less been introduced a few years later. Read the Swiss newspapers of the time when someone first suggested building mountain railways. What did they not throw at his head! Yet within a short time the mountain railways were there, and to-day no one remembers that he who devised them was a fool. Here, as in all things, it is simply a question of breaking down prejudice.

As I said before, if these two plants should be difficult to get in some locality, they might be replaced by something else, though it would certainly not be so good. Moreover, you can perfectly well use the plant as dried herb. On the other hand, most difficult to replace for its good influence on our manure is a plant which we are frequently not at all fond of—I mean, in the sense that you like to stroke what you are fond of. This is a plant we do not like to stroke—it is the *stinging nettle*. Truly it is the greatest benefactor of plant growth in general, and you will scarcely find another plant to replace it. If it should happen to be unobtainable in any place, then you must get it dried from elsewhere.

The stinging nettle is a regular " Jack-of-all-trades." It can do very, very much. It, too, carries within it the element which incorporates the Spiritual and assimilates it everywhere, namely, sulphur, the significance of which I have explained already. Moreover, the stinging nettle carries potassium and calcium in its currents and radiations, and in addition it has a kind of *iron* radiation. These iron radiations of the nettle are almost as beneficial to the whole course of Nature as our own iron radiations in our blood. Truly, the stinging nettle is such a good fellow and does not deserve the contempt with which we often look down on it where it grows wild in Nature. It should really grow around man's heart, for in the world outside —in its marvellous inner working and inner organisation—it is wonderfully similar to what the heart is in the human organism. The stinging nettle is the greatest boon.

Forgive me, Count Keyserlingk, if I become a little local in my references at this moment. But I would say, if ever it should be necessary in a certain sense to rid the soil of iron, you would do well to plant stinging nettles where they will do no harm. For in a certain sense the nettle plants would liberate the uppermost layers of the soil from the iron influence, because they are so fond of it and draw it into themselves. Though this might not undermine the iron as such, it would certainly undermine the influences of the iron on plant-growth in general. Hence it would undoubtedly be of great benefit to grow stinging nettles in this district. However, I only mention that in passing, to show you how important the mere presence of the stinging nettle may be for the growth of plants in the whole area around.

Now, to improve your manure still more, take any stinging nettles you can get, let them fade a little, press them together slightly, and use them in this case without any bladder or intestines. You simply bury the stuff in the earth. Add a slight layer of peat-moss or the like, so as to protect it from direct contact with the soil. Bury it straight in the earth, but take good note of the place, so that when you afterwards dig it out again you will not be digging out mere soil. There let it spend the winter and the following summer—it must be buried for a whole year.

This substantiality will now be extremely effective. Mix it with the manure, just as you did the other preparations. The general effect will be such that the manure becomes inwardly sensitive—truly sensitive and sentient, we might almost say intelligent. It will not suffer any undue decompositions to take place in it—any improper loss of nitrogen or the like.

This " condiment " will make the manure intelligent, nay, you will give it the faculty to make the earth itself intelligent—the earth into which the manure is worked. The soil will individualise itself in nice relationship to the particular plants which you are growing. It is like a permeation of the soil with reason and intelligence, which you can bring about by this addition of *Urtica dioica*.

What, after all, do they amount to—the customary modern methods of improving the manure? No doubt their first superficial effects are sometimes surprising, but the result will soon be that the alleged " excellent agricultural products " which you obtain thereby become mere stomach-filling for the human being. They will no longer have the proper nutritive power. You should not be deceived by the swollen size of any product. The point is that it should be inwardly consistent, with really nutritive intensity.

Now we may be concerned, here or there in our farming work, with the occurrence of *plant diseases*. I am speaking in general terms at the moment. Nowadays people are fond of specialisation in all things; therefore they speak of this disease or that. It is quite right to do so. If we pursue pure science, we must know what one thing or another looks like. Yet it is generally of little use for the doctor to be able to describe an illness ever so clearly. Far more important it is for him to be able to heal it, and in healing quite other points of view are important than those that the scientists generally have to-day in their description of diseases. We can attain the greatest perfection in the description of disease, we can know precisely what happens in the organism in terms of modern physiology or physiological chemistry; and yet we may still not be able to heal the disease at all. In healing we must proceed not from the histological or microscopic diagnosis, but from the great universal connections. And so it is in relation to plant-nature.

Moreover, plant-nature in this respect is simpler than animal or human nature; therefore our healing too can take—if I may say so—a more general course. For the plant world, we can indeed apply a kind of universal remedy. Indeed if it were not so, we should be in a very awkward position over against the vegetable world, as we often are over against the animals in veterinary work—of which, by the way, we shall still have to speak. This difficulty does not

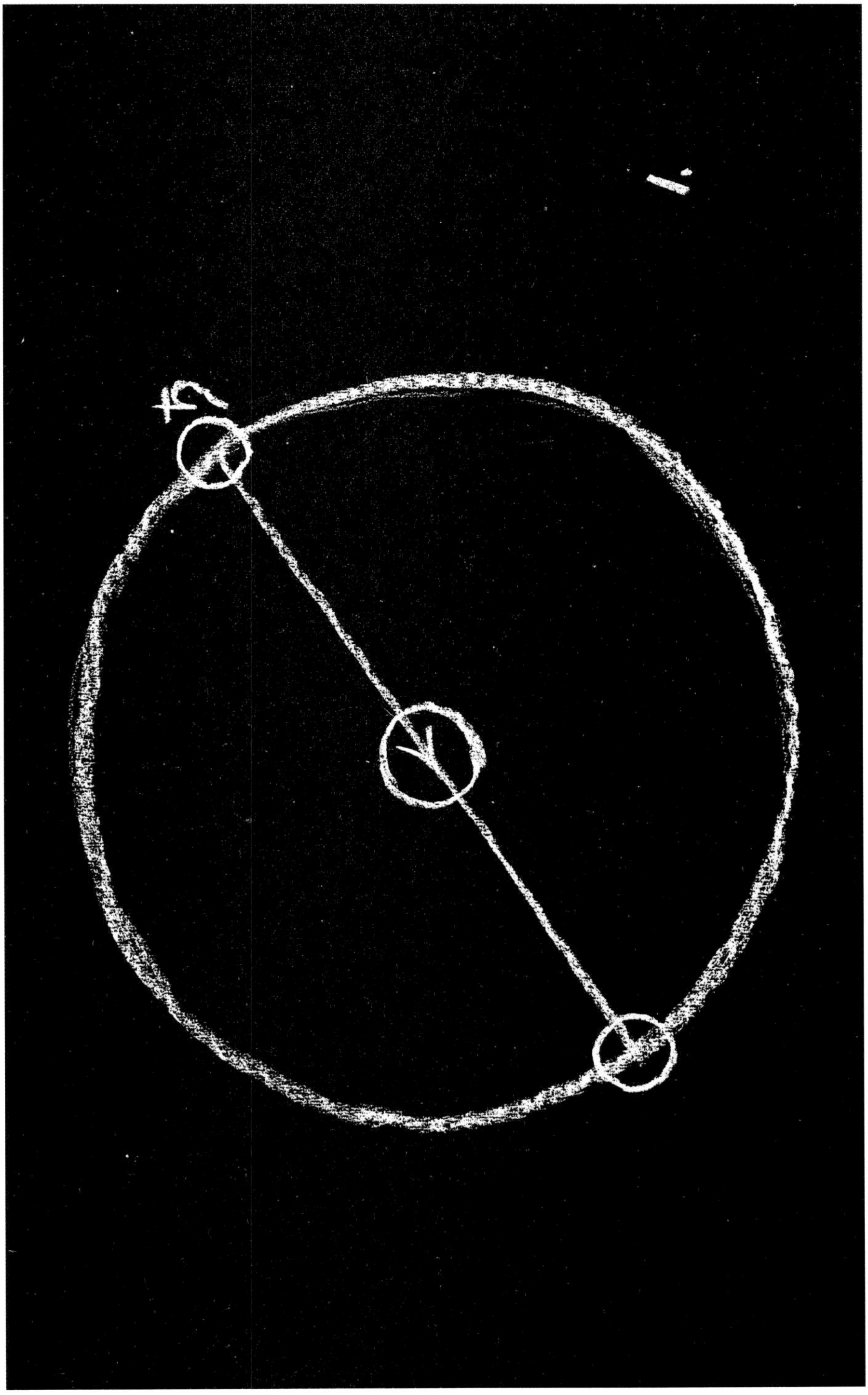

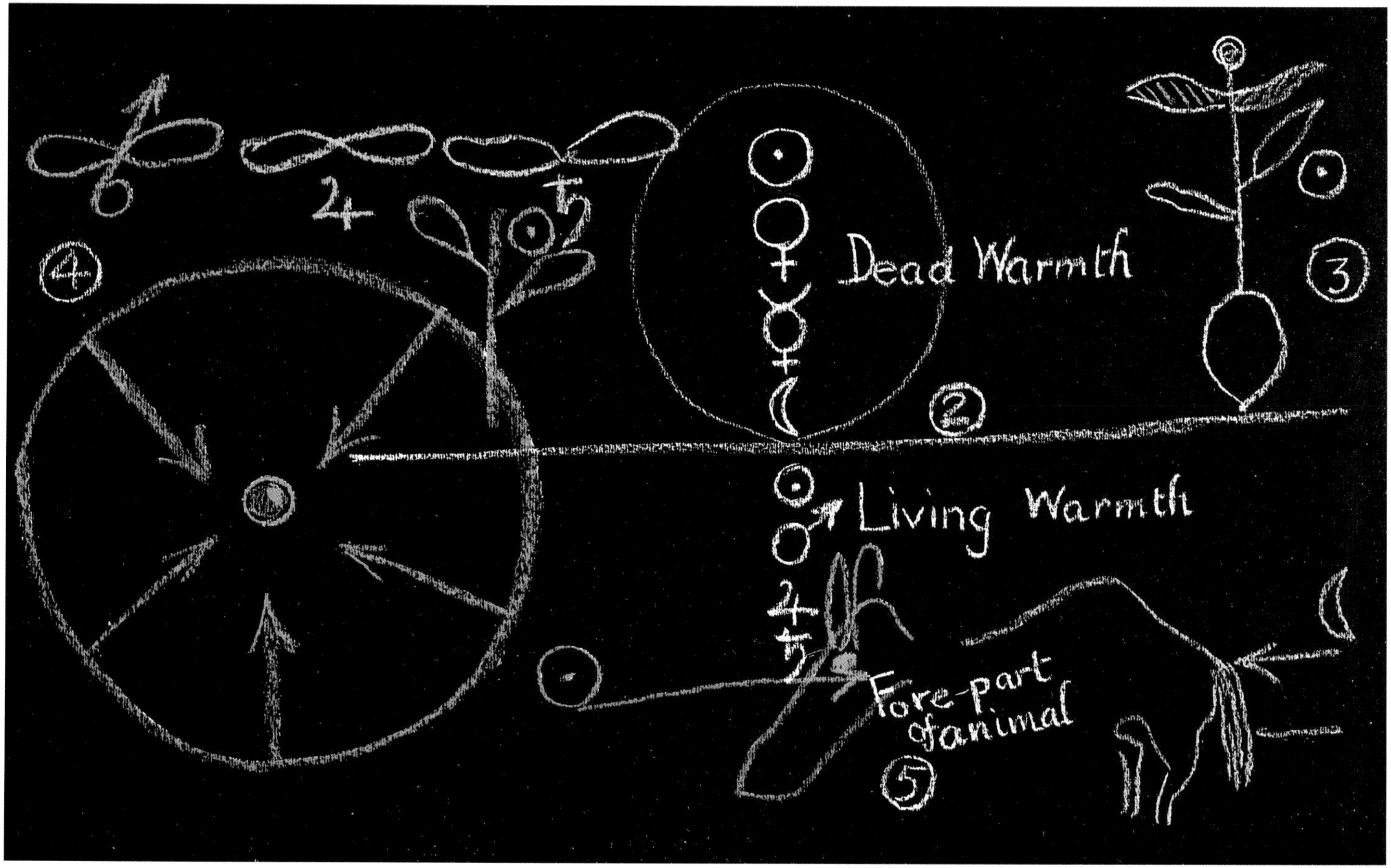

4
Dead Warmth
3
2
Living Warmth
Fore-part of animal
5

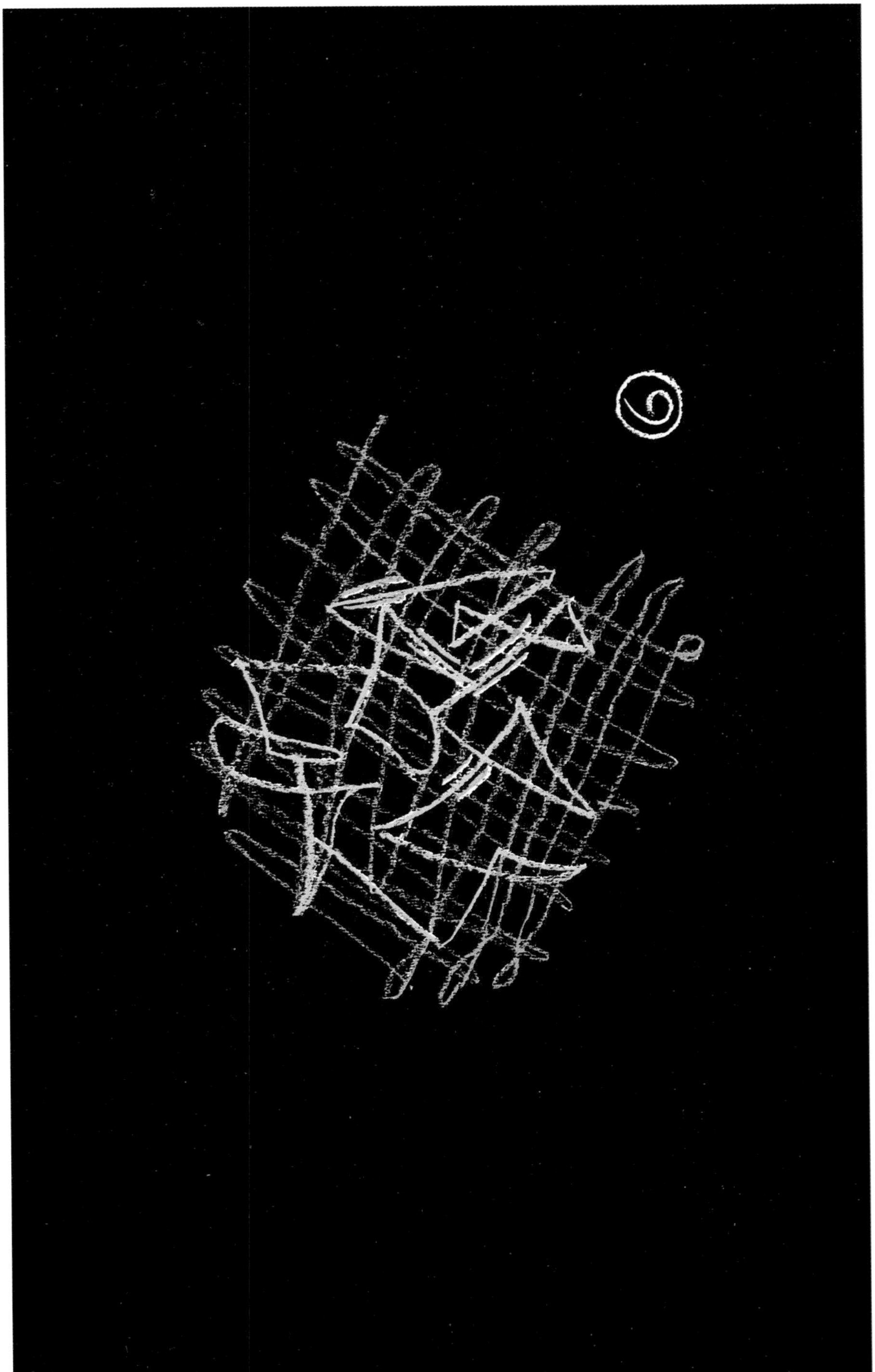

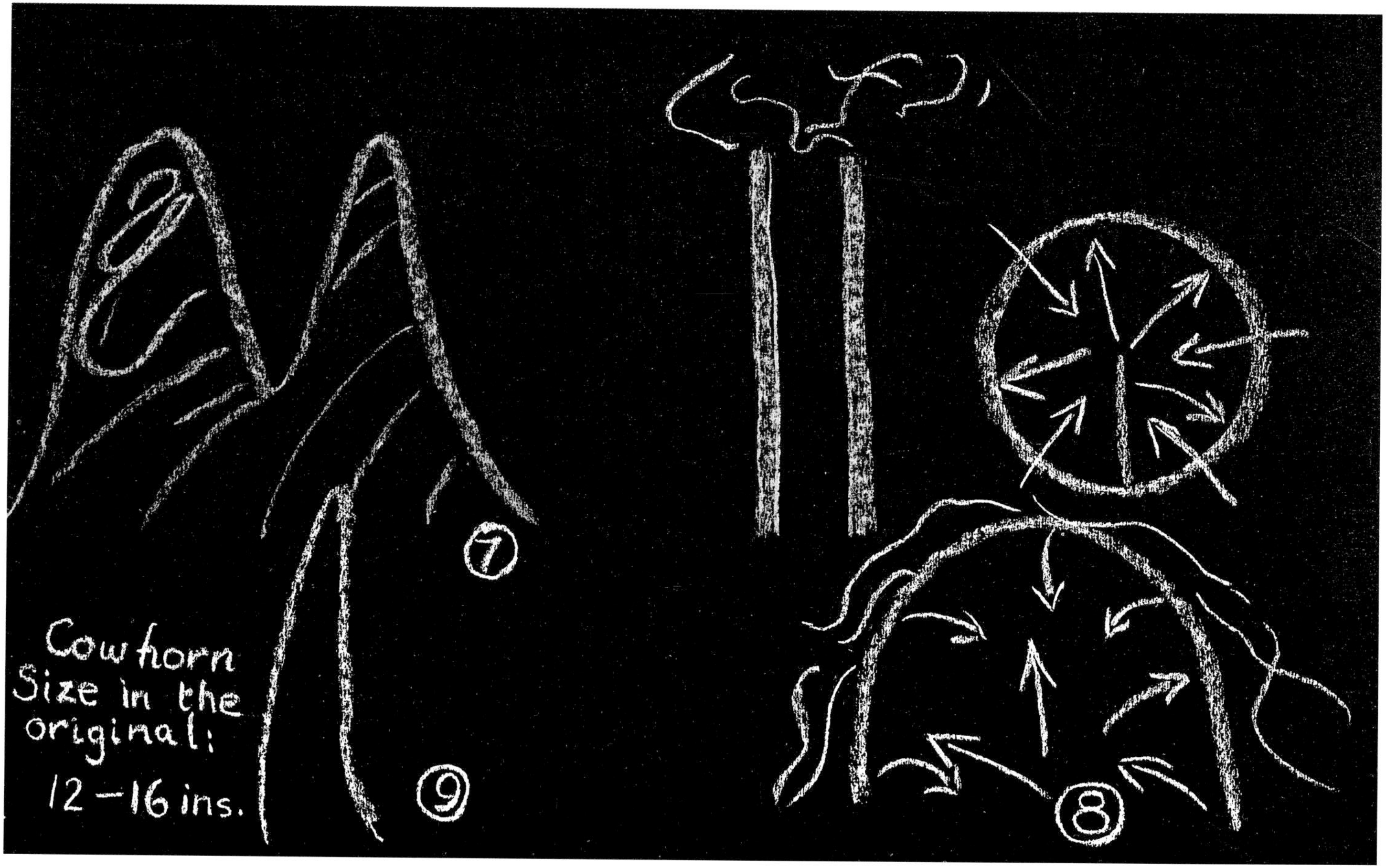

7
9
8
Cowhorn
Size in the
original:
12–16 ins.

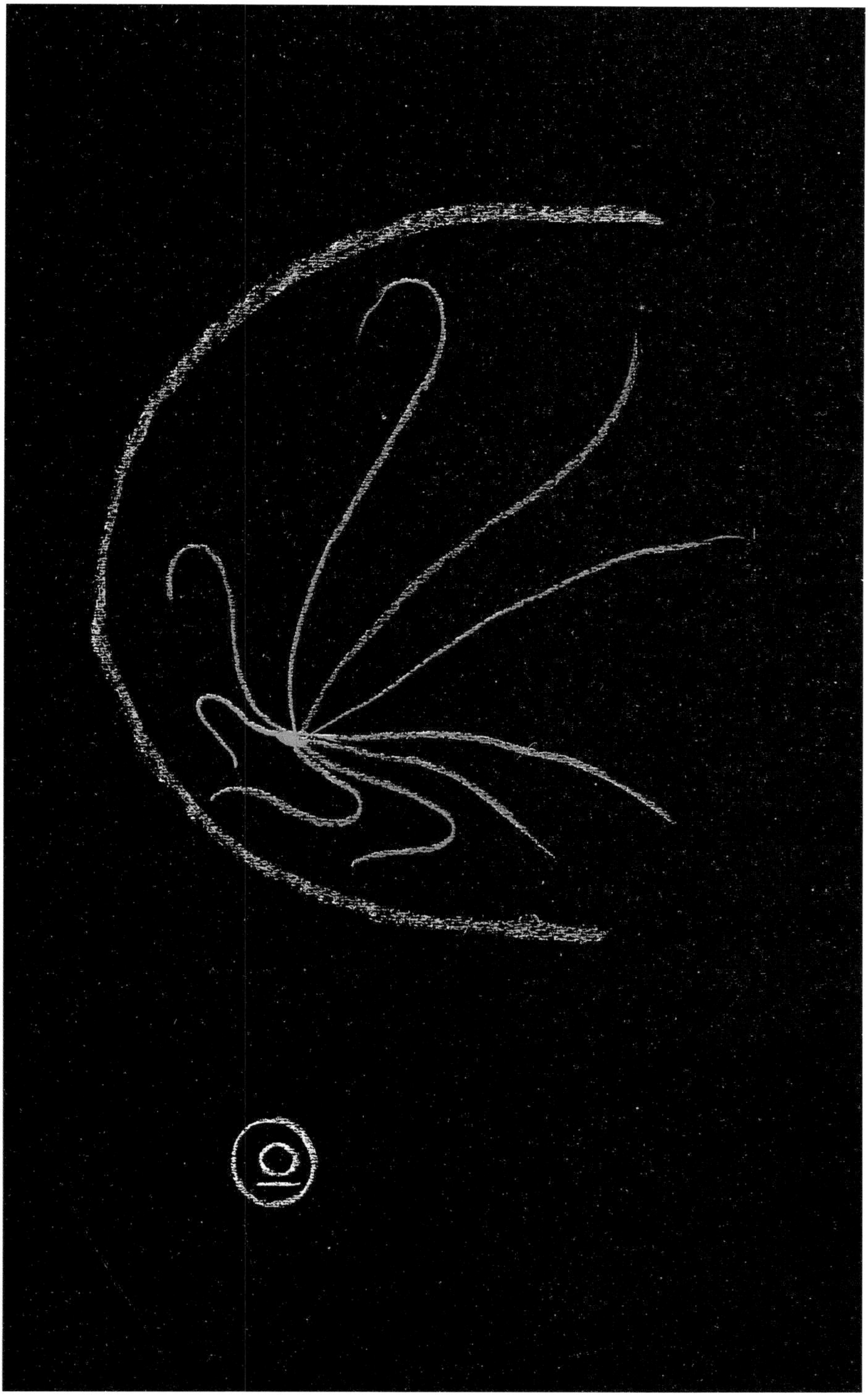

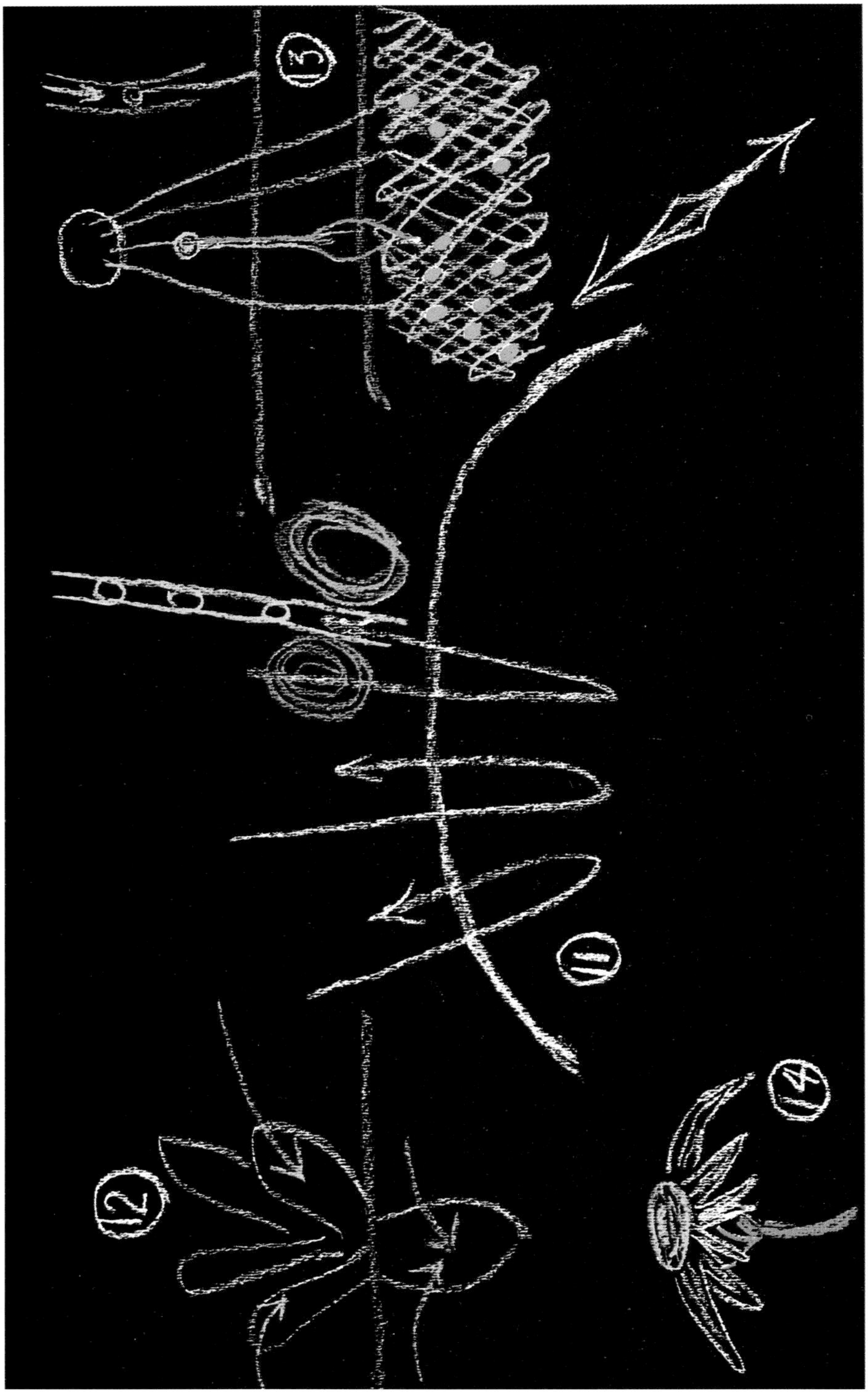

13
11
12
14

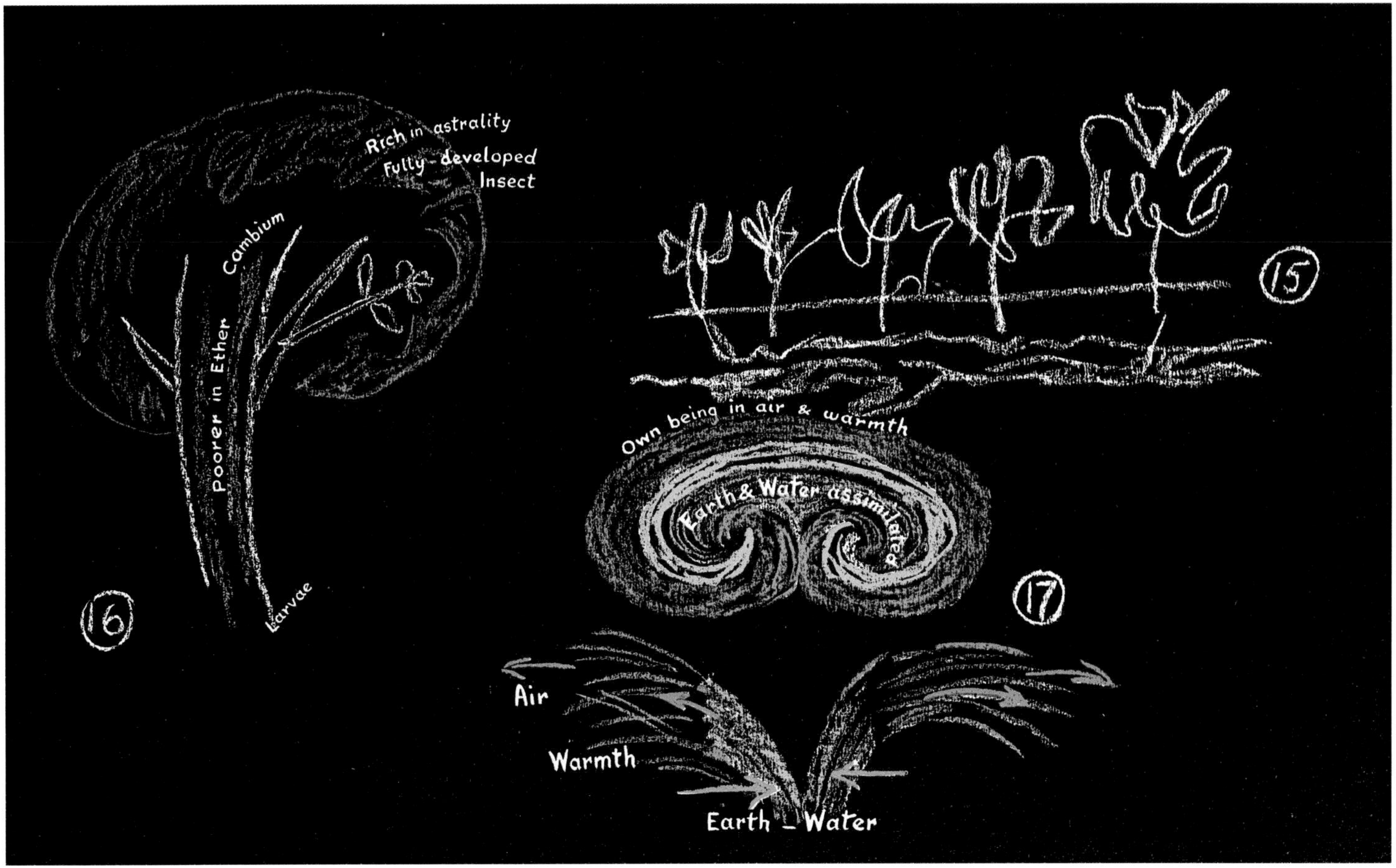
16
Rich in astrality
Fully-developed
Insect
Cambium
Poorer in Ether
Larvae
15
Own being in air & warmth
Earth & Water assimilated
17
Air
Warmth
Earth _ Water

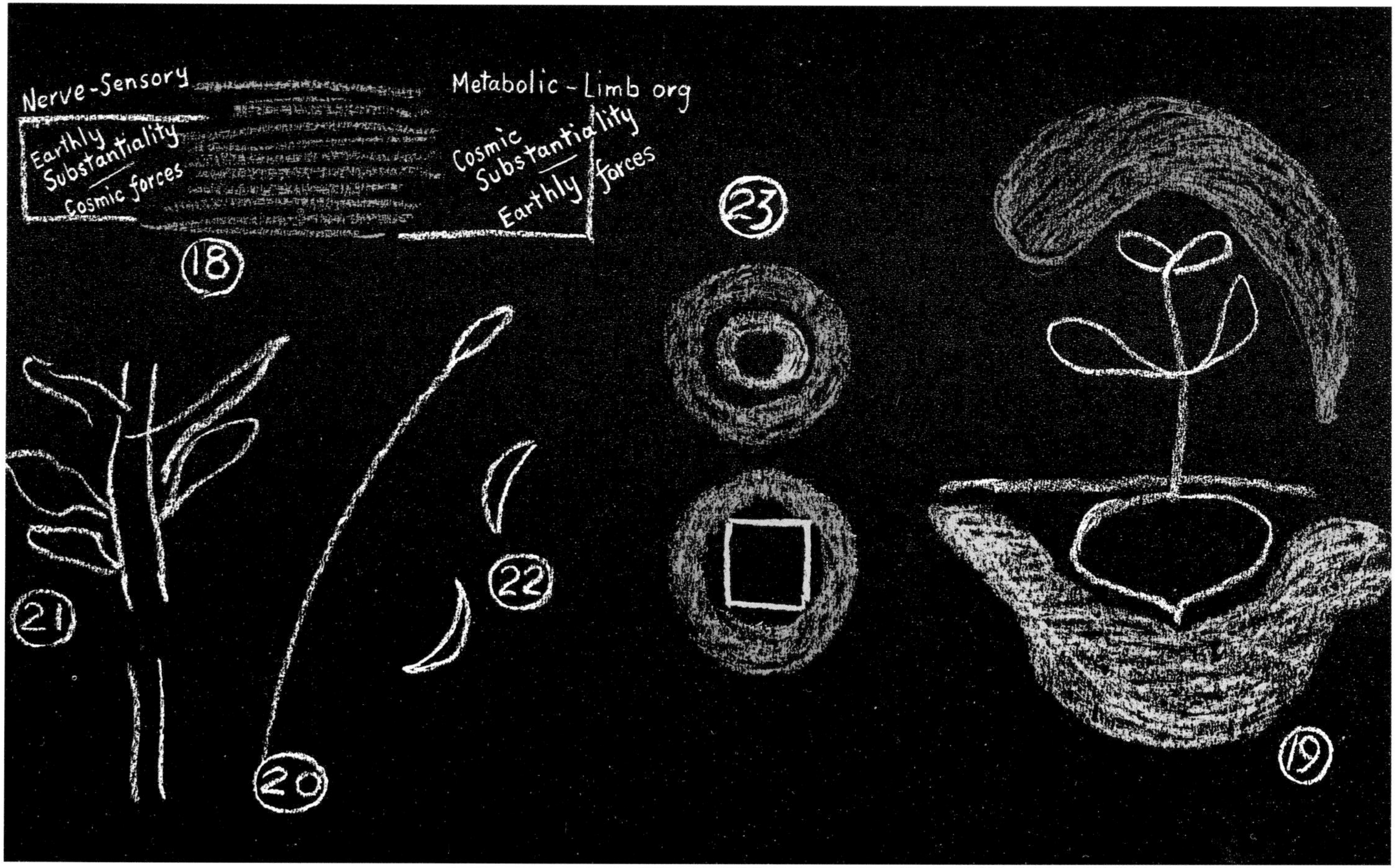
Nerve-Sensory
Earthly Substantiality
Cosmic forces
Metabolic - Limb org
Cosmic Substantiality
Earthly forces
18
19
20
21
22
23

occur in human healing, for a man can say what hurts him, while animals and plants can not. However, it is a fact that healing in this instance takes a more universal course. A large number of plant diseases, although not all, can be removed as soon as we observe them, by a rational improvement in our manuring, *i.e.* by the following methods.

We must bring calcium into the soil by our manure. But it will not be of use to bring the calcium to the soil by any channels that avoid the living sphere. To have a healing effect, the calcium must remain within the realm of life; it must not fall out of the living realm. Ordinary lime or the like is of no use at all in this respect.

Now there is a plant containing plenty of calcium—77 per cent of the plant substance, albeit in a very fine state of combination. I refer to the *oak*—notably the rind of the oak, which represents an intermediate product between plant-nature and the living earthy nature, quite in the way I explained when I spoke of the kinship of the living earth with bark or rind. For calcium as it appears in this connection, the calcium-structure in the rind of the oak is absolutely ideal.

Now calcium, when it is still in the living state, not in the dead (though even in the dead it is effective)—calcium has the property which I explained once before. It restores order when the ether-body is working too strongly, that is, when the astral cannot gain access to the organic entity. It " kills " or damps down the ether-body, and thereby makes free the influences of the astral body. So it is with all limestone. But if we want a rampant ethereal development, of whatsoever kind, to withdraw in a regular manner—so that its shrinking is beautiful and regular and does not give rise to shocks in the organic life—then we must use the calcium in the very structure in which we find it in the bark of the oak.

We collect oak-bark, such as we can get. We do not need much —no more than can easily be obtained. We collect it and chop it up a little, till it has a crumb-like consistency. Then we take a *skull* —the skull of any of our domestic animals will do, it makes little or no difference. We put the chopped-up oak-bark in the skull, close it up again as well as possible with bony material, and lower it into the earth, but not too deep. We cover it over with peat-moss, and then introduce some kind of channel or water-pipe so as to let as much rain-water as possible flow into the place. (We might even do it as follows: Take a barrel where rain-water is constantly flowing in and out. Put in it vegetable matter such as will bring about the continued presence of some vegetable slime. Let the bony vessel which contains the crumbled oak-bark lie in the slime in the water). This, once again, must hibernate. Snow-water is just as good as rain-water. It must pass through the autumn and winter in this way. What you add to your manuring matter from the resulting mass will lend it the forces, prophylactically to combat or to arrest any harmful plant diseases.

So we have added four different things. All this requires a certain amount of work, it is true—yet if you think it over, after all it involves less work than all the devices that are pursued in the chemical laboratories of modern agriculture, which are also costly. You will soon see that from the point of view of national economy what we have here explained pays better.

But we shall also need something to attract the silicic acid from the whole cosmic environment, for we must have this silicic acid in the plant. Precisely with regard to silicic acid, the Earth gradually loses its power in the course of time. It loses it very slowly, therefore we do not notice it. Nor must you forget that those who only look at the microcosmic or microscopic and never at the macrocosmic spheres, are unconcerned in any case about this loss of silicic acid; they think it insignificant for the growth of plants. In reality, it is of the greatest significance.

There is something you must know in this connection. For the scientists of to-day it will no longer argue such entire confusion on our part as it would have done a short time ago. Are not they themselves already speaking frankly of a transmutation of the elements? Observation of several elements has tamed the materialistic lion in this respect, if I may say so. Processes, however, that are taking place around us all the time are as yet utterly unknown. If they were known, people would more readily believe such things as I have just explained.

I know quite well, those who have studied academic agriculture from the modern point of view will say: " You have still not told us how to improve the nitrogen-content of the manure." On the contrary, I have been speaking of it all the time, namely, in speaking of yarrow, camomile and stinging nettle. For there is a hidden alchemy in the organic process. This hidden alchemy really transmutes the potash, for example, into nitrogen, provided only that the potash is working properly in the organic process. Nay more, it even transforms into nitrogen the limestone, the chalky nature, if it is working rightly.

You know that in the growth of plants, all the four elements of which I have been speaking are involved. Hydrogen also is there, in addition to sulphur. I have told you of the significance of hydrogen. Now there is a mutual and qualitative relationship between the limestone and the hydrogen, similar to that between oxygen and nitrogen in the air.

Even externally, in a quantitative chemical analysis as it were, the relationship between the oxygen-nitrogen connection in the air, and the limestone-hydrogen connection in the organic processes, might well be revealed. The fact is that under the influence of hydrogen, limestone and potash are constantly being transmuted into something very like nitrogen, and at length into actual nitrogen. And the nitrogen which is formed in this way is of the greatest benefit to plant-growth. We must enable it to be thus engendered by methods such as I have here described.

Silicic acid contains silicon as you know, and silicon, too, is transmuted in the living organism—transmuted into a substance of great importance, which, however, is not yet included among the chemical elements at all. Silicon is transmuted. In fine, we need the silicic acid to attract and draw in the cosmic properties. Now in the plant there simply must arise a clear and visible interaction between the silicic acid and the potassium—*not* the calcium. By the whole way in which we manure the soil, we must quicken it, so that the soil itself will aid in this relationship.

We must now look for a plant which by its own relationship between potassium and silicic acid can impart to the dung—once more, if added to it in a kind of homoeopathic dose—the corresponding power. And we can find it. This, too, is a plant which if it only grows among our farms, has a most beneficial influence in this direction. It is none other than the common *dandelion* (*taraxacum officinale*).

The innocent yellow dandelion! In whatever district it grows, it is the greatest boon; for it mediates between the silicic acid finely, homoeopathically distributed in the Cosmos, and that which is needed as silicic acid throughout the given district of the Earth. Truly this dandelion is a kind of messenger of Heaven. But if we need it especially—if we want to make it effective in the manure—we must use it in the right way. To this end—it will almost go without saying at this stage—we must expose the dandelion to the influences of the Earth, and in the winter season.

Here, too, we must gain the surrounding forces by a similar treatment as in the other cases. Gather the little yellow heads of the dandelion and let them fade a little. Press them together, sew them up in a *bovine mesentery*, and lay them in the earth throughout the winter.

In springtime you take the balls out, and you can keep them now until you need them. They are now thoroughly saturated with cosmic influences. The substance you get out of them can once again be added to the dung, and in a similar way. It will give the soil the faculty to attract just as much silicic acid from the atmosphere and from the Cosmos as the plants need, to make them really sentient to all that is at work in their environment. For they of themselves will then attract what they need.

To be able to grow truly, the plants must have a kind of sensation. Even as I, a human being, can pass a dull fellow by and he will not notice me, so too all that is in the soil and above it will pass a dull plant by, and the plant will fail to sense it; will not, therefore, enlist it in the service of its growth. But if the plant is thus finely permeated and vitalised with silicic acid, it will grow sensitive to all things, and will draw to itself all that it needs.

We can easily bring the plant into such a condition that it only needs a limited environment—immediately around it in the soil—to draw to itself what it needs. But it is not good to do so. Treat the soil of the earth as I have now described, and the plant will be prepared to draw things to itself from a wide circle. Your plant will then benefit not only by what is in the tilled field itself, whereon it grows, but also by that which is in the soil of the adjacent meadow, or of the neighbouring wood or forest. That is what happens, once it has thus become inwardly sensitive. We can bring about a wonderful interplay in Nature, by giving the plants the forces which tend to come to them through the dandelion in this way.

And so I think you should try to create good manures, by adding these five ingredients—or suitable substitutes—to your manuring matter in the way indicated. Manures in future should not be treated with all manner of chemicals, but with these five: yarrow, camomile, stinging-nettle, oak-bark and dandelion. Such a manure will have very much of what is actually needed.

Now you have one more river to cross. Before you make use of the manure thus prepared, press out the flowers of *Valerian.** Dilute the extract very highly. (You can do it at any time and keep it, especially if you use warm water in dilution). Add this diluted juice of the Valerian flower to the manure in very fine proportions. Then you will stimulate it to behave in the right way in relation to what we call the " phosphoric " substance.

With the help of these six ingredients you can produce an excellent manure—whether from liquid manure, or ordinary farmyard-manure, or compost.

* *Valeriana officinalis.*

DISCUSSION

KOBERWITZ,
13th June, 1924.

Question: When you speak of the bladder of the stag, are you referring to the male animal?

Answer: Yes.

Question: Do you mean the annual or the perennial nettle?

Answer: Urtica dioica.

Question: Is it right to roof in the manure-pit in districts where there is much rain?

Answer: The manure ought to be able to stand any ordinary amount of rain. It is not good for it to get no rain-water at all. On the other hand, it should not be thoroughly washed out with rain; that, of course, would harm it. You cannot decide by hard-and-fast rules. Generally speaking, rain-water is good for manure.

Question: Should not the place where the manure is stored be walled-in and covered over to prevent the loss of the manure-juice?

Answer: In a certain sense, the manure *needs* rain-water. The only thing is, it might sometimes be well to keep the rain off a little by spreading granulated peat over the top. There is no purpose in keeping the rain away altogether by roofing it in. That would undoubtedly deteriorate the manure.

Question: If plant-growth is stimulated to such an extent by the manuring methods you have indicated, are cultivated plants and so-called weeds equally stimulated? Must any special methods be adopted to destroy the weeds?

Answer: In the first place the question is justified, needless to say, and I shall speak of the combatting of weeds in the next few days. What I have given you so far is favourable to plant-growth in general; you would not thereby put an end to the growth of weeds. On the other hand, it will make the plants far more secure against any parasitic pests that might occur. Here you have already the remedy against such parasitic pests as may occur in the plant kingdom. The combatting of weeds, on the other hand, does not arise out of the principles which we have hitherto discussed. The weed naturally shares in the general plant-growth. We shall yet have to speak on this subject. The whole thing is so intimately connected that it would not be well to pick out any special aspect now.

Question: What do you hold of the method of Captain Krantz? By piling it up in loose layers, and taking advantage of the spontaneous generation of warmth, the manure is also made odourless.

Answer: I have purposely refrained from speaking of what is already being done on rational lines. I wanted to give the inspirations which can come from Spiritual Science for the improvement of every such method. The one you refer to has many advantages, no doubt, but I believe it is comparatively new; it is not a very old method. And it may be this is also one of the methods which appear a dazzling success to begin with, but do not prove quite so practical in course of time. When the soil has its tradition, so to speak, everything will in a way refresh it; but when you apply the same method for a longer time, it is often as it is in medicine. When a medicament comes into the body for the first time, why, the most unbelievable medicaments are helpful the first time you take them! But then the curative effect is at an end. Here too it always takes some time before you recognise that it is not as you were first led to believe.

The one thing of importance is the spontaneous generation of warmth. The activity that must come into play for the generation of this warmth is exceedingly good for the manure; of that there can be no doubt. This activity cannot but lead to good results. Possible disadvantages might arise from the manure being piled up loosely; nor do I know if it is quite literally true, as you suggest, that it becomes quite odourless. If you *do* really get it odourless, it would indicate that the method is really good and beneficial. I believe it has not been tried for many years.

Question: Is it not better to pile up the manure *above* the earth than to sink it in a pit below the level of the ground?

Answer: In principle it is generally right to put it as high as possible. You should not, however, put it *too* high; you must still keep it in proper relation to the forces that are there beneath the earth. You cannot actually put it on a hillock, but you can build it up from the normal level of the ground; that will give you the most favourable height.

Question: Can the same compost methods be applied to the vine, which has suffered so much in recent times?

Answer: Yes, but with modifications. I shall mention some modifications when I come to speak of fruit- and vine-growing. Generally speaking, what I have given to-day applies to the improvement of every kind of manure. I have indicated what will improve manure in general. The specific modifications of these methods for meadow- and pasture-land, cereal crops, orchards and vineyards still remain to be dealt with.

Question: Is it right to have the manure-ground paved or plastered?

Answer: From all that one can know of the whole structure of the earth and its relation to the manure, it would be utterly wrong. I cannot see why it should be paved. If your manure-ground is paved or plastered, you should hollow out a space all around so as to leave room for the interplay of the manure with the earth. Why deteriorate the manure by separating it from the earth?

Question: Has the ground beneath it any influence—whether, for instance, it be sandy or clayey? Sometimes the ground layer of the place where the manure is to be kept is covered with clay so as to make it impervious.

Answer: Undoubtedly the different kinds of earth will have their influence, according to their specific properties *as* kinds of earth. If there is sandy ground where you want to store the manure, it will be necessary to fill it in with a little clay. For the sand is pervious and will suck in the water. If, on the other hand, you have a very clayey soil, you should loosen it a little, and sprinkle in some sand. For a medium effect, always take a layer of sand and a layer of clay. Then you have both—the inner consistency of the earth-kingdom and also the watery influences. Otherwise the water will trickle away. A mixture of the two kinds of earth will be the best. For the same reason you should not choose a ground of " Loess " to pile up your manure-heap—not if you can avoid it. " Loess," or the like, will not be very helpful. In such a case it will be better to create in course of time an artificial ground for your manure-heap.

Question: As to the cultivation of the plants you mentioned—yarrow, camomile, the stinging nettle—could they be introduced into a district by scattering the seed, if they did not happen to be growing there already? In cattle-farming we have generally assumed that yarrow and dandelion too are dangerous for cattle. We therefore wanted to exterminate these plants as far as possible—likewise the thistle. Indeed we are now engaged in doing so. I presume we should now have to sow them again along the edges of the fields, but not in the meadows and pastures?

Question by Dr. Steiner: But how should they be harmful as animal food?

Count Keyserlingk: Yarrow is said to contain poisonous substances. Dandelion is said to be not good for cattle.

Dr. Steiner: You should watch it carefully. On the open field, an animal will not eat it if it is really harmful.

Count Lerchenfeld: We in our district do the very opposite. We treat the dandelion as good fodder for milch cattle.

Dr. Steiner: These are sometimes mere prevalent opinions; nobody knows if they have ever been tested. It is possible, no doubt, that in the hay . . .—it would have to be tested—I think, if it were harmful, an animal would leave the hay untouched. An animal will not eat what is not good for it.

Question: Has not yarrow largely been removed by the large doses of lime? Yarrow surely needs a moist and acid soil?

Answer: If you use wild yarrow, a very small quantity will suffice, even for a large estate. It has a peculiar, homoeopathic effect. If you had some yarrow in the garden here, it would be enough for the whole estate.

Question: I for my part have observed that the young dandelion, shortly before flowering, is very gladly eaten by all cattle. Afterwards, however, when it has begun to blossom, the cattle will no longer take it.

Answer: You must always remember the following: this, at least, is the general rule. An animal will not eat dandelion if it is harmful. An animal's feeding instinct is excellent.

You must also bear this in mind. We too, when we wish to stimulate something that depends on a living process, will almost always use what we should not use by itself. For instance, no one would eat yeast as his daily food; yet it is used in baking bread. A thing that even can act as a poison when consumed in large doses will, under other conditions, have the most beneficial effects. After all, medicines are generally poisonous.

The *process*—not the substance—is important. Thus I believe you can well get over your misgivings about the dandelions doing harm to your animals. So many strange ideas are prevalent. It is curious: here, on the one hand, the harmfulness of the dandelion is emphasised by Count Keyserlingk, while on the other hand, Count Lerchenfeld describes it as the best of milch-fodder. The effects cannot possibly be so different in two such neighbouring countries; one or another of the two opinions must be wrong.

Question: Perhaps it is a question of the underlying basis? My statement was founded on veterinary opinions. Ought we then purposely to plant yarrow and dandelion on our pasture and meadowland?

Answer: Quite a small surface will suffice.

Question: Does it depend on how long the preparations are kept with the manure, after taking them out of the earth?

Answer: Once they are mixed with the manure it is meaningless to ask how long they should be kept in it. But it should all have been done *before* the manure is spread over the fields.

Question: Should the manure-preparations be put into the earth all together, or each one separately.

Answer: That is of some importance. While the interaction is going on, the one preparation should not be allowed to disturb the other. Therefore it is well to dig them in some distance apart. If I had to do it on a small estate, I should dig them in as far as possible from one another, so as to prevent their interfering with each other. I should look for the most distant parts around the edge of the estate. On a large estate you can choose the distances as you will.

Question: Does it matter if the earth above the preparations is overgrown, once they are buried?

Answer: The earth can do as it likes. It is quite good if it is grown over. It may even be overgrown with cultivated plants.

Question: How should the preparations be dealt with *in* the manure-heap?

Answer: I should advise the following procedure. Prick a hole about a foot deep, or a little deeper, in a large pile of manure, so that the manure can close up again around the stuff. You need not make it as deep as a metre, but the manure ought to be able to close up again round the preparations. For it is like this (Diagram 10):

If this is the pile of manure, and you have here a little of the preparation . . . it all depends on the radiations. The rays go out like this; it is not well if the stuff is too near the surface. The radiation is thrown back from the surface; it returns in a definite curve. It does not go outside, provided the manure closes up around the substance. Half a metre (about 18 inches) will suffice. If it is too near the surface, a considerable portion of the rays of force will be lost.

Question: Is it enough if you only make a very few holes, or should the preparations be distributed as widely as possible?

Answer: It is better to distribute them—not to make all the holes in one place. Otherwise the radiations may interfere with each other.

Question: Should all the preparations be put into the manure at the same time?

Answer: When you are putting the preparations in the manure-heap, you can put in the one beside the other. They do not influence each other; they only influence the manure as such.

Question: Can the preparations all be put into one hole?

Answer: Theoretically, even if all the preparations were put into one hole, one might presume that they would not disturb each other; but I should not like to make this statement *a priori*. You can put them in fairly close together, but they might after all interfere with each other, if you mixed them all up in a single hole.

Question: What kind of oak did you mean?

Answer: Quercus robur.

Question: Must the bark be taken from a living tree, or will a felled tree do?

Answer: As far as possible from a living tree; nay, more, from a tree in which you may presume that the " oak resin " is still pretty active.

Question: Is it the whole of the bark?

Answer: No, only the surface—the outermost layer of bark which crumbles off of its own accord when you loosen it.

Question: In burying the manure preparations, is it absolutely necessary to go no deeper than the fertile layer? Or could one bury the cow-horns even deeper?

Answer: It is better to leave them in the fertile layer. Indeed it may be presumed that in the subsoil underneath the fertile layer they would no longer provide fruitful material. You should, however, consider that the best possible condition would be provided by a layer of fertile soil as deep as you can find. Look for a place where the fertile layer is deepest—that will undoubtedly be the best. Beneath the fertile layer you will get no beneficial effect.

Question: Within the fertile layer they will always be exposed to the frost. Will that do no harm?

Answer: If exposed to the frost, they come into the very time when the earth, by virtue of the frost, is most intensely exposed to cosmic influences.

Question: How should you grind down the quartz or the silica? In a small grinding-mill, or in a mortar?

Answer: In this case the best thing will be to do it first in a mortar; and you will need an iron pestle. Grind it down in the mortar to a fine, mealy consistency. If it is quartz, having ground it down as far as possible in this way, you will even need to continue grinding it afterwards on a glass surface. It must be a very fine meal, and that is not easy to attain with quartz.

Question: Farming experience shows that a well-nourished head of cattle puts on substance which was lacking. There must therefore be a relation between the actual feeding and the absorption of nutritive substance from the atmosphere?

Answer: You need only observe what I said. In the absorption of food, the *forces* developed by the body are the essential thing. Thus it depends on the receiving of proper food, whether or no the animal develops sufficient forces to be able to receive and assimilate the substances from the atmosphere.

You may compare it with this: If you have a very close-fitting glove to put on, you cannot do it by sheer force. You wedge the glove out with a wooden instrument; you thus extend and stretch it. So too in this case; the forces have to be made pliant and supple. Such forces must first be there, for the creature to receive from the atmosphere what it does not get from the actual food. The food is there to stretch the organism, so to speak, thus enabling it to receive all the more from the atmosphere. This may even lead to hypertrophy if too much is taken, and you would pay for it by the shorter duration of the creature's life. There is a happy mean here, too, between the maximum and minimum.

LECTURE SIX

KOBERWITZ,
14th June, 1924.

MY DEAR FRIENDS,

The further course of our studies must be based on such insight as we have already gained into plant-growth, and into animal formations too. Aphoristically at least, we must now consider a few among the spiritual-scientific ideas that relate to harmful plants and animals and to what are commonly called plant diseases. These things can only be studied in concrete detail. Very little can be said in general terms; they must all be specifically dealt with. Therefore, to begin with, I will give examples which—taken as the starting-point for your experiments—will lead you on to further instances.

First let me deal with weeds and harmful plants in general. We are not so much concerned to define " weeds." We only want an insight into the problem, how to rid a given field or area of plants which we do not want to have there. You know, one sometimes has strange harkings-back to one's student days. Thus I endeavoured, though with no great enthusiasm, to look up a few text-books to see how they defined " the weed." Most of the authors, I found, if they tried to define what a weed is, described it thus: " Everything that grows at a place where you do not want it is a weed "—a definition which certainly does not take us very far into the essence of the matter.

Indeed, we shall have little good fortune in considering the essence of " weeds " as such—for the simple reason that in Nature's judgment a weed has just as much right to grow as a plant which we find useful. These things must be looked at from a somewhat different point of view. The simple question is, how can we rid a certain field or area of what will naturally grow there through the prevailing conditions of Nature, while *we* do not want it there?

We can only answer this question by taking into account what we have dealt with in the past few days. I showed how we must strictly distinguish between the forces that are there in the growth of plants—forces which, though they come from the Cosmos, are first received into the earth and then work from the earth upon plant-growth. As I said, the forces which are mainly due to the cosmic influences of *Mercury*, *Venus* and *Moon* (though they do not work directly from these planets, but by the round-about way of the Earth)—these are the forces we must consider when we are tracing what produces the daughter-plant after the mother-plant and so on in succession. While on the other hand, in all that the plant derives from the surrounding sphere, from that which is over the earth, we must perceive the workings and potentialities which the more distant planets transmit to the air, which are in this way received.

Moreover, speaking in a wider sense we may say: All the forces that work into the earth from the near planets are influenced by the chalk-or *limestone*-workings of the earth, while that which works from the surrounding sphere is influenced by the workings of *silica*. Although the silica influences proceed from the earth itself, nevertheless they transmit what proceeds originally from *Jupiter*, *Mars* and *Saturn*—not what proceeds from Moon, Venus and Mercury. Nowadays, people are altogether unaccustomed to take these things into account. They pay the penalty for their ignorance. Indeed, in many regions of the civilised world a heavy penalty has been paid for this ignorance of the cosmic influences—ignorance both of the cosmic influence when it works through the air through all that lies above the level of the ground, and of the cosmic influence when it works from below through the mediation of the earth. They have had to pay the penalty for this lack of insight.

It happened in widespread regions of civilisation. (It may be of no concern to you, but it is a very grave concern for many people). They had exhausted all the resources that were once upon a time applied. They had exhausted all that had been done since ancient times by an old instinctive science. Not only the soil of the Earth was exhausted—the traditions too were exhausted, though sometimes simple peasant folk would lend a helping hand. So it has come about: far and wide, the *vine plantations* have been subjected to the ravages of the grape-louse,* and they are pretty helpless against it. I could tell you a tale of the editorial offices of a Viennese agricultural paper in the 1880's. They were approached from every side to find a remedy against the grape-louse, and they were at a loss. For by that time the plague had grown acute. These things cannot be treated thoroughly by the scientific methods of to-day. They can only be dealt with effectively by entering into all that can be known along the lines which we have indicated here.

Let me show it diagrammatically (Diagram 11). Imagine this as the level of the earth's surface. Here we have all the influences that come in from the Cosmos—from Venus, Mercury and Moon—and ray back again, working upward from below. Everything that works in the earth in this way causes the plants to bring forth what grows in a single year and culminates in seed-formation. From the seed a new plant arises, and a third, and so on. Once more then: everything that works from the Cosmos in this way flows out into the reproductive forces—into the sequence of generations.

On the other hand there is all that which comes by another way—above the level of the earth—all that which comes from the forces of the distant planets. Diagrammatically we can draw it thus: it represents all that is transformed in the plant so that it spreads out and expands in the surrounding circle. Here therefore we have what makes the plant look thick or bulky—*i.e.* what we can take away as nourishment, because a continuous stream reforms it, ever

* *Phylloxera vastatrix.*

anew. I mean, for example, what we take from the apple- or the peach-tree—the fleshy fruit which we consume. All this is due to the influences of the *distant planets*.

Such insight alone will tell us how to act if we wish to influence the plant's growth in a particular way. It is only by taking these varied forces into account that we gain an idea, how we can influence the plant's growth. Now a large number of plants—notably those which we ordinarily count as weeds—are greatly influenced by the workings of the Moon. These are often medicinal plants. Precisely among the " weeds," so-called, we often find the strongest curative herbs.

What do we know of the Moon in ordinary life? We know that it receives the rays of the Sun upon its surface and throws them back again on to the earth. We see the rays of the Sun reflected—we catch them with our eyes—and the Earth, too, of course, receives these rays from the Moon. It is the rays of the Sun which are thus reflected, but of course the Moon permeates them with its own forces. They come to the Earth as lunar forces, and so they have done ever since the Moon separated from the Earth.

Now in the Cosmos it is just this lunar force which strengthens and intensifies all that is earthly. Indeed, when the Moon was united with the Earth, the Earth itself was far more living, fruiting, inherently fertile. When the Moon was still one with the Earth there was nothing so mineral as we have to-day. Even now, after its severance, the Moon works so as to intensify the normal vitality of the Earth, which is still just enough to bring about the growth in living creatures. The Moon intensifies it, thus enhancing the growth-process to the point of reproduction.

Whenever a being grows, it becomes larger. In this process the very same force is at work as in reproduction. Only in growth it does not go so far as to bring forth a fresh being of the same species. It brings forth cell upon cell. That is a feebler reproductive process—one that remains within the limits of the single entity. What we commonly call reproduction is an enhanced growth-process.

Now the Earth by itself is still just able to transmit that feeble reproductive process which growth represents; but it has no power, without the Moon's assistance, to produce the enhanced growth-process of reproduction. Here it requires the cosmic forces shining in upon the Earth through the Moon—and, in the case of certain plants, through Mercury and Venus too. As I said, people commonly imagine that the Moon merely receives the Sun's rays and throws them down on to the Earth. In considering the Moon's effect they only think of the Sunlight; but that is not the only thing that comes to the Earth.

With the Moon's rays the whole reflected Cosmos comes on to the Earth. All influences that pour on to the Moon are rayed back again. Thus the whole starry Heavens—though we may not be able to prove it by the customary physical methods of to-day—are in a sense rayed back on to the Earth by the Moon. It is indeed a strong and powerfully organising cosmic force which the Moon rays down into the plant, so that the seeding process of the plant may also be assisted; so that the force of *growth* may be *enhanced* into the force of *reproduction.*

However, all this is only there for a given district of the Earth when it is full Moon. When it is new Moon, the country does not enjoy the benefit of the Moon-influences. It only holds fast in the plants, during the new Moon, what they received at the full Moon. Indeed, we should attain important results if we only tried to see what progress we could make by using the Moon, let us say, in sowing—*i.e.* for the very earliest germinating activity within the Earth. So the old Indians used to do until the nineteenth century. They also sowed according to the phases of the Moon.

However, Nature is not so cruel as to punish man forthwith for his slight inattention and discourtesy to the Moon in sowing and in reaping. We have the full Moon twelve times a year, and that is adequate for a sufficiency of the full-Moon influences, *i.e.* of the forces that quicken the fruiting process. If on any occasion we perform what tends to fertilisation, not at the full Moon but at the new, it will simply wait in the Earth till the next full Moon. So it gets over our human errors and takes its cue from great Nature.

This is sufficient for men to make use of the Moon all unawares. But that is all—and we get no farther along these lines. Treated in this way, the weeds will demand their rights just as much as the vegetables, and everything grows confused, for we are strangers to the forces that regulate growth. We must first enter into them. Then we shall know that by using the fully evolved Moon-force we work for the reproduction of all vegetable life, *i.e.* for that which shoots up from the root, right up into the seed-formation. Thus we shall get the strongest of weeds if we let the kind Moon work down upon them—if we do nothing to arrest its influence upon our weeds. For there are wet years when the Moon-forces work more than in the dry. The weeds will then reproduce themselves and increase greatly.

If on the other hand, we reckon with these cosmic forces, then we shall say to ourselves: We must contrive to check the full influence of the Moon upon the weeds. That is to say, we must only let work upon them the influences coming from without—not the Moon-influences, but those that work directly. Then we shall set a limit to the propagation of the weeds; they will be unable to reproduce themselves. Now we cannot " switch off " the Moon. Therefore we must treat the soil in such a way that the earth is disinclined to receive the lunar influences. Indeed, not only the earth, but the plants, too (*i.e.* the weeds) can thus become disinclined to receive the lunar influences. We can make the weeds reluctant, in a sense, to grow in earth which has thus been treated. If we attain this end, we have all that we need.

You see the weeds growing rampant in a given year. You must accept the fact. Do not be alarmed; say to yourself: Something must now be done. So now you gather a number of seeds of the weed in question. For in the seed the force of which I have just spoken has reached its final culmination. Now light a flame—a simple wood-flame is best—and burn the seeds. Carefully gather all the resulting ash. You get comparatively little ash, but that does not matter. Quite literally, for the plants thus treated by letting their seeds pass through the fire and turn to ash, you will have concentrated in the ash the very opposite force to that which is developed in attracting the Moon-forces.

Now use the tiny amount of substance you have thus prepared from a variety of weeds, and scatter it over your fields. You need not take especial care in doing so, for these things work in a wide circumference. Already in the second year you will see, there is far less of the kind of weed you have thus treated. It no longer grows as rampantly. Moreover, many things in Nature being subject to a cycle of four years, after the fourth year you will see, if you continue sprinkling the pepper year by year, the weed will have ceased to exist on the field in question. Here, in fact, you will make fruitful the " effects of smallest entities," which have now been scientifically proven in our Biological Institute.

Much might be attained in this way. Quite generally speaking, you have far-reaching possibilities if you really reckon on these influences which remain unconsidered nowadays. Thus, for the dandelion which you need as I explained yesterday, you can perfectly well plant it where you want it, and use the dandelion-seed. Repeat this fire-process with it, prepare your little pepper and scatter it over the fields. Then you will have the dandelions where you want them, and at the same time keep the fields, thus treated with burnt dandelion, free of the dandelion plant.

People to-day will not believe it ; such things were known and mastered once upon a time by an instinctive farming wisdom. They could plant together, in circumscribed areas, whatever they wanted to have. They knew of these things instinctively.

In all these matters, I can only give indications, but as you see, these indications are capable of direct practical application. And as there is still the prevailing judgment—I will not call it prejudice—that all things must be subsequently verified, good and well! Set to work and try to verify them. If you do the experiments rightly, you will soon see them confirmed. If I had a farm, however, I should not wait to see them verified. I should apply the method at once, for I am sure that it will work. So it is for me. Spiritual-scientific truths are true in themselves, we need not have them confirmed by other circumstances or by external methods.

Our scientists have all made this mistake of looking to external methods to verify these truths. In the Anthroposophical Society, too, our scientists have done so. They at least should have known better;

they should have known that a thing can be true in itself. However, to get anywhere nowadays we must always verify things externally. It is no doubt a necessary compromise; in principle it is not necessary. One knows of these things inwardly. They stand inherently, by their own quality—that is how one knows them.

To take another illustration. Suppose I have something manufactured by fifty workers. I say to myself: I want to produce three times as much, therefore I will employ 150. Now comes a clever fellow and declares, I do not believe that 150 workers will produce three times as much; you must first put it to the test. Let us suppose you make the experiment. You get your work done—whatever it may be—first by one, then by two and then by three people, and now you tell statistically how much the three get done between them. Well, if so be they spent their time in chattering, they may have done even less than the one worker. Your premiss is wrong; your experiment has proved the opposite. But it proves nothing in reality. If you are working exactly, you must consider the other case with equal exactitude. If you do so, whatever is inherently true will beyond doubt be outwardly confirmed.

Thus we can speak, more in general terms, of the harmful plants or vegetable pests of the field. But we can no longer speak so generally when we come to the animal pests. Let me choose one example—a characteristic instance, whereon you can make your experiments and see how these things are confirmed in practice.

There is a very good friend of the farmer—the field-mouse. What do they not try to do to fight against it! Read of it in the agricultural text-books. To begin with, all manner of phosphorus preparations were used; then, other things, such as the " Strychnine-Saccharine " preparations. Nay, an even more radical method has been proposed, namely, to infect the field-mice with typhus. Certain bacilli, harmful only to rodents, are added to mashed potatoes and the bait is distributed. Such things have also been done—at least, they have been recommended.

So they try to get at these happy, simple-looking little creatures in untold ways—by methods which do not look very humane, to say the least. They try to attack the mice once they are there. I think even the State is being set in motion. When you attack the mice in this way, it is no good unless the neighbouring farmer also does so, for they only come back from the neighbouring field; and so the State must be called in to see that everyone is compelled to drive the mice away by standard methods. The State will have no modifications. It makes its regulations once for all. Once it has judged a method right—no matter whether it is so or not—it decrees that everyone must do it. It issues general regulations.

All these are mere external rulings and experiments at random, and one has an underlying feeling: the experimenters themselves are not quite happy about it. For in the end the mice always come back again. What we need to do in this case is also not quite applicable on a single estate by itself, though to some extent it may help even then. It will not be very easy to carry out. One will have to work towards a general insight, so that one's neighbours too will do it. (I venture to say that in the future we must look far more to intelligent insight than to police regulations. That will be progress in our social life).

And now, imagine that you do the following: You catch a fairly young mouse and skin it, so as to get the skin. There you have the skin of a fairly young mouse. (There are always enough mice—albeit, they must be *field*-mice if you wish to make this experiment). But you must obtain this skin of the field-mouse at a time when Venus is in the sign of Scorpio.

Those people of olden time, you see, were not so stupid with their instinctive science! Now that we are passing from plants to animals, we come to the " animal circle "—that is, the " Zodiac." It was not called so in a meaningless way. To attain our end within the plant-world we can stop at the planetary system. For the animal world, that is not enough. There we need ideas that reckon with the surrounding sphere of the fixed stars, notably the fixed stars of the Zodiac.

Moreover, in the growth of plants the Moon-influence is well-nigh sufficient to bring about the reproductive process. In the animal kingdom, on the other hand, the Moon-influence must be supported by that of Venus. Nay, for the animal kingdom the Moon-influence does not need to be considered very much. For the animal kingdom conserves the lunar forces; it emancipates itself from the Moon. The Moon-force is developed in the animal kingdom even when it does not happen to be full Moon. The animal carries the force of the full Moon within it, conserves it, and so emancipates itself from limitations of time.

This does not apply to what we here have to do; it does not apply to the other planetary forces. For you must do something quite definite with the mouse-skin. At the time when Venus is in Scorpio, you obtain the skin of the mouse and burn it. Carefully collect the ash and the other constituents that remain over from the burning. It will not be much, but if you have a number of mice, it is enough. You can easily get enough.

Thus you obtain your burned mouse-skin at the time when Venus is in Scorpio. And there remains, in what is thus destroyed by the fire, the corresponding negative force as against the reproductive power of the field-mouse. Take the pepper you get in this way, and sprinkle it over your fields. In some districts it may be difficult to carry out; then you can afford to do it even more homoeopathically; you do not need a whole plateful.

Provided it has been led through the fire at the high conjunction of Venus and Scorpio, you will find this an excellent remedy. Henceforth, your mice will avoid the field. No doubt they are cheeky little beasts; they will soon come out again if the pepper has been so sprinkled that a few areas remain unpeppered in the neighbourhood. There they will settle down again. Undoubtedly the influence of it rays out far and wide; nevertheless, it may not have been done quite thoroughly. But the effect will certainly be radical if the same is done in the whole neighbourhood.

I venture to think that you will have considerable pleasure in such things. You may begin to find your farming very tasty—like certain dishes are when they have been a little peppered. So we begin really to reckon with the influences of the stars without becoming superstitious in the least. Many things afterwards became mere superstition, which were originally knowledge. You cannot warm-up the old superstitions. You must make a fresh start with genuine knowledge. This knowledge, however, must be gained in a spiritual way—not through the mere physical world-of-the-senses.

This is the way to treat the earth, if you have to combat field-vermin which can be reckoned in any sense among the higher animals. Mice are rodents; they are included among the higher animals. But you will not do much with the insects in this way. Insects are subject to different cosmic influences. Indeed, all the lower animals are subject to different cosmic influences than the higher animals. And now for once allow me to tread upon thin ice and mention the *nematode* of the root crops as an example; so you will have something near at hand.

The so-called " beginning " of the disease is seen in the well-known swellings of the rootlet and in the limpness of the leaves in the morning. That is the external sign. Now we must remember that this middle part (it is the leaves that here suffer a change) absorbs the cosmic influences from the air; whereas the roots absorb those forces which come into the plants from the cosmos *via* the Earth.

What happens now, when the nematode appears? The absorption of cosmic forces which should normally be going on in the region of the leaves is pressed downward, into a region where it eventually comes near to the roots. Diagrammatically speaking, we may say (Diagram 12), if this be the surface of the earth, and this the plant, then—in the nematode-infested plant—the cosmic forces which should be working up above are working down here below. This is the real phenomenon. Certain cosmic forces are sliding too far down. Hence, too, the outward appearance of the plant. But this too gives the animal the power to receive within the earth, where it must live, the cosmic forces upon which its life depends. For it would otherwise have to be living in the leaves. (The nematode is a wire-like worm). But it cannot live up there, for the earth is its natural domain.

Some living creatures, nay, all living creatures have this peculiarity: they can only live within certain limits of existence. You try to live in an air whose temperature is seventy degrees centigrade, hot or cold, above or below zero. You cannot do it. You depend on a certain temperature. Above and beneath this level you can no longer live. Nor can the nematode. It cannot live if the earth is not there, nor can it live unless the cosmic forces are there at the same time. Otherwise it would have to die out. Thus, for each living creature, there are quite definite conditions. The human race too would die out if it were not for certain conditions.

Now for the creatures that evolve in this particular way, it is important for the cosmic element which normally makes itself felt only in the Earth's surrounding sphere, to come right down *into* the Earth. Moreover, these influences take place in periods of four years. The nematode is something highly abnormal. To recognise its nature, we might equally well investigate the cockchafer-grubs which come in cycles of four years. The forces are the same in both cases. The very same forces which give the Earth the tendency to unfold the potato-seedling—these forces the Earth also receives for the formation of the cockchafer-grubs, which occur with the potatoes every four years. Wherever this is so, we have a four years' cycle. Though it does not apply to the nematode itself, it certainly applies to what we must do in counteracting it.

In this case you do not take part of the insect as you do with the mouse. You must take the entire insect. An insect like this, which settles harmfully in the plant-root, is altogether an outcome of cosmic influences; it only needs the Earth as its underlying basis. Therefore you must burn the whole insect. It is best to burn it; that is the quickest way. You might also let it decay; possibly this would be even more thorough, only it is difficult to collect the products of decay. But you will certainly attain what you need by burning the whole insect.

Now it is necessary to perform this operation when the Sun is in the sign of Taurus. (If need be, you can keep the insect and burn it when the time comes). This, you see, is precisely the opposite of the constellation in which Venus must be when you prepare your mouse-skin pepper. In effect, the insect world is connected with the forces that evolve when the Sun is passing through Aquarius, Pisces, Aries and Gemini and on to Cancer. In Cancer it appears quite feebly, and it is feeble again when you come to Aquarius. It is while passing through these regions that the Sun rays out the forces which relate to the insect world.

People are unaware what a specialised thing the Sun is. The Sun is not really the same when in the course of a year or a day it shines on to the Earth from Taurus, or from Cancer, or the other constellations. In each case it is different. It is comparative nonsense to speak of the Sun in general terms—albeit, pardonable nonsense. We should really speak of Aries-Sun, Taurus-Sun, Cancer-Sun, Leo-Sun, and so on. For the Sun is a different being in each case. Moreover, the resultant influence depends both on the daily course and on the yearly course of the Sun, as determined by its position in the vernal point.

If you do this—if you thus prepare your insect-pepper—once

again you can spread it out over the beet-fields, and the nematode will by and by grow faint—a faintness you will certainly find very effective after the fourth year. For by that time the nematode can no longer live. It shuns life if it has to live in an earth thus peppered.

In a strange way we come again to what was formerly described as " Wisdom of the Stars." Modern astronomy serves as a mere mathematical orientation, nor can we put it to any other use. It was not so in former ages. Time was when they saw in the stars something from which they could take their direction for earthly life and work. Such science is utterly lost to-day.

In this way, therefore, we can also hold the animal pests at bay. It is important for us to come into relation to the Earth in this way. We must be aware of these things. On the one hand, it is right that the Earth should receive the faculty to bring forth plant-life out of itself. This faculty the Earth receives, as we have seen, mainly through the Moon- and watery influences. But that which is in the plant—nay, that which is in every living being—also carries within it the seed of its own annihilation.

Just as water on the one hand is a *sine qua non* of all fertility, so on the other hand, fire is an absolute destroyer of fertility. Fire consumes fertility. Therefore, if you treat by fire in the proper way that which is normally treated by water to bring about fertility in the plant-world, you will bring about destruction—annihilation in the household of Nature. These are the things you must consider. A seed will develop fertility far and wide through the Moon-saturated water; likewise a seed will develop forces of annihilation far and wide through the Moon-saturated fire—and altogether, through the cosmically-saturated fire, as we have seen in the last example.

After all, our reckoning upon this great force of dispersal (while pointing out the precise effects of time in the process) need not seem utterly strange to you. The force of the seed always works in dispersal and expansion. Hence, in the force of annihilation too, it works far and wide. Expansive power lies inherent in seed-nature. It is the very property of the seed to have this power of dispersal; so, too, the pepper we prepare in this way has a real expansive power. (I only call it pepper on account of its appearance. The preparations generally look like pepper).

It only remains for us to consider so-called *plant diseases.* Properly speaking, we cannot really say " plant diseases." The rather abnormal processes which occur as plant-diseases are not diseases in the same sense as in animal diseases. (We shall understand the difference more exactly when we come to the animal kingdom). Notably, they are not at all the same kind of process as in human diseases.

Properly speaking, disease is not possible without the presence of an astral body. In an animal or human being, the astral body is connected with the physical through the ethereal. There is a certain normal condition. The astral body may be connected *more* intensely with the physical (or with any one of its organs) than it should normally be. In such a case, the ether-body fails to provide a sufficient cushioning or "padding," and the astral body drives into the physical too strongly. It is under these conditions that most of our illnesses arise.

Now the plant has in it no real astral body. Hence the specific way of being ill, which can occur in the animal and in the human being, does not occur in the plant. We must be well aware of this fact. Thus we must first gain an insight into the question, what is it that can bring about illness of plants?

You will have seen, from my descriptions, how the whole earth in the plant's environment has an inherent life of its own. With all this life in the Earth—albeit not so intensely as to bring forth plant-forms, yet nevertheless with some intensity—manifold forces of growth and faint suggestions of reproductive forces are present all around the plant. Moreover, there is all that which is working in the Earth under the influence of the full-Moon forces, mediated by the water. Here is a wealth of significant relationships.

You have the Earth—the Earth which is filled with water—and you have the Moon. The Moon, letting its radiations pour into the Earth, makes it to some extent alive in itself; awakens waves and weavings of the ethereal within the Earth. It does so more easily when the earth is saturated with water, and with greater difficulty when the earth is dry. You must remember, the water is only a mediator. It is the earth itself—the solid, mineral element—which must be made alive. The water, too, is mineral. There is of course no hard-and-fast line. Thus we must have the lunar influences in the soil.

Now the Moon-influences in the soil can also become too strong. This can happen in a very simple way. You need only call to mind a thoroughly wet winter, followed by a thoroughly wet spring. Then the Moon-forces will enter the earth too strongly. The earth will become too much alive. Once more, you will have an over-intense vitalisation of the earth. I will indicate it by making little red dots (Diagram 13) where the earth is too strongly vitalised by the Moon. If the little red dots were not there—if the earth were not over-vitalised by the Moon—plant-life would grow upon it, developing normally up to the seed: corn, for instance, growing upward to the seed.

If the Moon imparts precisely the right vitality to the earth, this vitality will work on and upward till the seed develops. Assume now that the Moon-influence is *too* strong; the earth is too much vitalised. Then it will work too strongly from below upward. That which should only occur in the seed-formation will occur at an earlier stage. Precisely when it is too strong, it will be insufficient to reach to the top. Through its very intensity, it will work itself out more in the lower regions. As a result of the strong Moon-influence, the seed-formation proper will have insufficient power.

The seed receives something of dying life into itself, and through this dying life there arises, as it were, above the soil—above the primary level of the earth—a secondary level. Although it is not earth, the same effects are there—above the proper level—and, as a consequence, the seed (the upper part of the plant) becomes a kind of soil for other organisms. Parasites and fungoid growths arise—all manner of fungoid growths.

Thus we see the forming of mildew, blight, rust, and similar diseases. The over-intense Moon-influence prevents what should work upward from the earth from reaching the necessary level. The true force of fertility depends upon the Moon's influence being normal. It must not be *too* intense. It may seem strange, but it is so: this result is brought about, not by a weakening but by an over-intensity of the Moon-forces. If we merely theorised about it instead of looking at the process, we might reach the opposite conclusion, but we should be wrong. Perception shows it as I have now described it. What, then, should we do?

We must somehow relieve the earth of the excessive Moon-force that is in it. And we *can* do so. We need only perceive what works in the earth so as to deprive the water of its mediating power; so as to lend the earth more " earthiness " and prevent it from absorbing the excessive Moon-influences through the water it contains. We can achieve this result. Outwardly, it all remains just as it is. But we now prepare a kind of tea or decoction—a pretty concentrated decoction of *equisetum arvense*.* This we dilute, and sprinkle it as liquid manure over the fields, wherever we need it—wherever we want to combat rust or similar plant-diseases. Here again, very small quantities are sufficient—a homoeopathic dose is quite enough.

Once more you see how the several fields of life work into one another. Understand the strange influence which equisetum arvense has upon the human organism through the function of the kidneys, and you will have your guiding line. Needless to say, you cannot merely speculate. Nevertheless, you have a guiding line, and you will now investigate how equisetum works when you transform it as described, into a kind of liquid manure, and sprinkle it over the fields. You need no special apparatus. It will work far and wide, even if you only sprinkle a very little, and you will find it an excellent remedy. Strictly speaking, it is not a medicament, for in the true sense of the word a plant cannot be diseased. It is not a healing process in the proper sense; it is simply the opposite process to the one I described.

So you must learn to see into the workings of Nature in all her different domains. Then you will really take the processes of growth in hand. (We shall afterwards see the same for animal growth—animal normalities and abnormalities). To get the growth-processes in hand—that is the really important thing. To experiment at random on these matters, as is done to-day, is no real science. The mere jotting-down of isolated notes and facts—that is no science.

* Mare's-tail, horse-tail, shave-grass.

Real science only arises when you begin to control the working forces. But the living plants and animals—even the parasites in the plants—can never be understood by themselves. What I said in our first lesson when I referred to the magnet-needle is only too true. Anyone who thought of the magnet-needle alone—anyone who looked in the magnet-needle itself for the causes of its always turning northward—would be talking nonsense. We do not do so; on the contrary, we take the whole Earth and assign to it a magnetic North Pole and a magnetic South. The whole Earth must be included in our explanation.

Just as we draw in the whole Earth to understand the properties of the magnet-needle, so, when we come to the living plants, we must not merely look at the plant or animal or human world; we must summon all the Universe into our counsels! Life always proceeds from the entire Universe—not only out of what the Earth provides. Nature is a great totality; forces are working from everywhere. He alone can understand Nature who has an open sense for the manifest working of her forces.

What does science do nowadays? It takes a little plate and lays a preparation on it, carefully separates it off and peers into it, shutting off on every side whatever might be working into it. We call it a "microscope." It is the very opposite of what we should do to gain a relationship to the wide spaces. No longer content to shut ourselves off in a room, we shut ourselves off in this microscope-tube from all the glory of the world. Nothing must now remain but what we focus in our field of vision.

By and by it has come to this: scientists always have recourse, more or less, to their microscope. We, however, must find our way out again into the macrocosm. Then we shall once more begin to understand Nature—and other things too.

DISCUSSION

KOBERWITZ,
14th June, 1924.

Question: Can the method given for the nematode be applied to other insects? I mean, to any kind of vermin? Is it permissible without further scruples to destroy animal and plant life in this way over wide areas? The method might be greatly abused. Some limit ought surely to be set, to prevent a man from spreading destruction over the world.

Answer: As to its being permissible, let us assume for a moment that such a thing were not permitted. (For the moment I will not speak of the ethical—occultly ethical—question). If such procedures were not allowed, what I have repeatedly hinted at would inevitably follow: agriculture would go from bad to worse in civilised countries. Not only intermittent periods of local starvation or high prices would occur, but these conditions would become quite general. Such a state of affairs may well be with us in a none too distant future. We have thus no other choice. Either we must let civilisation go to rack and ruin on the earth, or we must endeavour to shape things in such a way as to bring forth a new fertility. For our needs to-day, we really have no choice to stop and discuss whether or no such things are permissible.

Nevertheless, from another point of view, the question may still be asked; and from this aspect we should rather consider how to establish once more a kind of safety-valve against misuse. It goes without saying that when these things are generally known and applied, abuses will be possible; that is quite evident. Nevertheless, it may be pointed out that there have been epochs of civilisation on the earth when such things were known and applied in the widest sense. Yet it was possible for those among mankind who were in earnest to keep these things within such bounds that the misuse did not occur.

Abuses did indeed occur in an epoch when far graver abuses were still possible, because these forces were universally prevalent. I mean during the later periods of Atlantean evolution, when a far greater misuse occurred, leading to grave catastrophes. Generally speaking, we can only say that the custom of keeping the knowledge of these things in small circles and not allowing it to become more general, is justified; but in our times it is scarcely possible any longer. In our time knowledge cannot be retained in limited circles; such circles immediately tend in one way or another to let the knowledge out.

So long as the art of printing did not exist, it was easier; and at a time when most people were unable to write, it was easier still.

Nowadays, for practically every lecture—however small the circle where you hold it—the question is immediately raised: Where shall we get a shorthand writer? I do not like to see the shorthand writer; one has to put up with him, but it would be better if he were not there (I mean the shorthand writer, not the person, needless to say).

Must we not also reckon, on the other hand, with a further necessity—namely, the moral improvement of all human life? That alone can be the panacea against abuses—the moral upliftment of human life as a whole. Admittedly, when we consider certain phenomena of our time, we might become a little pessimistic; but in regard to this question of the moral improvement of life we should never tend to a mere contemplation of facts. We should always try to have thoughts that are permeated with impulses of *will*. We should consider what we can really *do* for the moral betterment of human life in general. This can arise from Spiritual Science. Spiritual Science will have nothing against it if a Circle is formed which will act from the outset as a means of healing against possible abuses.

After all, in Nature too it is so: everything good can become harmful. Think for a moment: if we had not the Moon-forces below, we could also not have them above. They simply must be there; they must be working. That which is requisite and necessary in one sphere in the highest degree, is harmful in another. That which is moral on one level is decidedly immoral on another. That which is Ahrimanic in the earthly sphere is only harmful because it is *in* the earthly sphere. When it takes place in a realm that is but a little higher, its effect is definitely good.

As to your other question, it is quite right: the method I indicated for the nematode applies to the insect world in general. It applies to all that portion of the animal world which is characterised by the possession of an abdominal marrow and not a spinal marrow. Where there is spinal marrow, you must first skin the animal. In the other case, the whole creature should be burned.

Question: Did you mean the wild camomile?

Answer: This camomile, with the petals turned downwards. (As in the drawing, Diagram 14.) It is the " Chamomilla officinalis " —growing wild by the wayside.

Question: Do you also take the flower of the stinging-nettle?

Answer: Yes, and you can take the leaves too—the whole plant at the time when it is flowering—only not the root.

Question: Can one also take the dog camomile that occurs in the fields?

Answer: That is a species more akin to the right one than the garden camomile which is now being shewn. The latter is quite useless. The one you refer to is also sometimes used for camomile tea. It is far more akin to the right one, and may be used if need be.

Question: I take it the camomile growing here along the railway track is the right one?

Answer: Yes, that is the right one.

Question: Will what you said of the destruction of weeds apply also to water-weeds?

Answer: Yes, it applies also to plants that grow out of the swamp or out of the water; it applies to water-weeds. In such a case you must sprinkle the banks with the pepper.

Question: Can underground parasites, as, for instance, the cabbage root-fly, be combatted by the same means?

Answer: Undoubtedly.

Question: Can the remedy for plant-diseases also be applied to the vine?

Answer: It has not yet been tested—I, too, have not tested it—and little has been done in this direction occultly. I can only say, I am convinced the vine could have been protected if one had gone about it in the way I have indicated.

Question: What of the so-called grape leaf-fall disease or downy mildew (*Plasmopara viticola*)?

Answer: It can be combatted in the same way as any other kind of rust, mildew or blight.

Question: Is it legitimate for us as anthroposophists to resuscitate vine-growing?

Answer: To-day, in many respects, Anthroposophical Science can only be there to say what *is*. The question of what *ought* to be is more difficult as yet, for many spheres of life. I knew a good anthroposophist friend who possessed extensive vineyards. However, he used a considerable portion—not all too large a portion—of his annual profits to send out postcards through the world preaching abstinence. On the other side, I had a friend who was himself a strict abstainer, and who, moreover, was very generous to the anthroposophical movement throughout his life. He was, however, responsible for the placards you see everywhere on the tramcars—" Sternberger Cabinett " (a kind of champagne). Here, then, the practical question becomes rather ticklish. You cannot get all you want nowadays. Therefore I said, it is the cow-horns which we take from the cows to bury in the earth. As to the bulls' horns which we might don, to run up against all and sundry in a bull-at-the-gate fashion—by so doing we might easily cause harm to Spiritual Science.

Question: Might not the bladder of the stag be replaced by something else?

Answer: No doubt it may be difficult to get stags' bladders; and yet—how many things that are difficult are not done in the world! One might of course try if one could not replace the bladder of the stag by something else; I cannot say at the moment. Maybe there is a species of animal somewhere—indigenous, perhaps, to some very limited territory in Australia for instance; but I can imagine nothing similar among the European native animals. In any case it would have to be an animal bladder. I cannot recommend you immediately to think of finding substitutes.

Question: Must the position of the stars always be the same for combatting insect pests?

Answer: It will have to be tested. I said that the whole series is important from Aquarius to Cancer. Undoubtedly, within these limits, a variation among the constellations for the different kinds of lower animals will be significant. It must be tested.

Question: Did you mean the astronomical Venus, for the field-mice?

Answer: Yes, that which we call the evening star.

Question: What " constellation of Venus with Scorpio?"

Answer: Whenever Venus is visible in the sky with the Scorpio constellation in the background. Venus must be behind the Sun.

Question: Has the burning of potato haulms any influence on the thriving of the potatoes?

Answer: The influence is so slight as to be practically negligible. There is indeed an influence; there is always a certain influence, whatever you do with any organic relic. It influences not only the single plants, but the entire field. But the influence is so small as to be practically negligible.

Question: What do you mean by " Rindergekröse " (bovine mesentery in Lecture 4)?

Answer: The peritoneum (" Bauchfell "). That surely is the generally accepted meaning of " Gekröse."

Question: Is it the same as " Kuttelflecke " (tripe)?

Answer: No, it is not the same. The peritoneum is meant.

Question: How should the ash be distributed over the fields?

Answer: I said just what I meant. You do it as though you were sprinkling pepper into something. It has so great a radius of influence that it is quite sufficient if you simply walk over the fields and sprinkle it.

Question: Do the preparations work in the same way on fruit-trees?

Answer: Generally speaking, all that I have said applies to fruit-culture also. A few things, still to be considered, will be given to-morrow.

Question: It is the custom in farming to give the farmyard manure to turnips and the like. Is the specially prepared manure important for cereals also, or should the latter be treated differently?

Answer: Existing customs can surely be retained, at any rate to begin with. The point is simply to *add* what I have indicated. As to other usages of which I have not spoken, you surely need not begin by representing everything as bad—trying to reform everything. Truly, I think you can continue the methods that have proved good, and supplement them with what has been given. I should, however, state that the influence of the methods I have indicated will be considerably modified if you use manure that is rich in sheep or pig dung. The effect will not be so striking as it will be if you avoid using sheep and pig dung to excess.

Question: What if one uses inorganic manures?

Answer: Mineral manuring is a thing that must cease altogether in time, for the effect of every kind of mineral manure, after a time, is that the products grown on the fields thus treated lose their nutritive value. It is an absolutely general law.

Precisely the methods I have given, if properly followed, will make it unnecessary to manure oftener than every three years. Possibly you may only have to manure every four or six years. You will be able to dispense with artificial manuring altogether. You will do without it if only for the reason that you will find it much cheaper to apply these methods. Artificial manure is a thing you will no longer need; it will go out of use.

Nowadays, opinions are based on far too short periods of time. In a recent discussion on bee-keeping, a modern bee-keeper was especially keen on the commercial breeding of queens. Queens are sold in all directions nowadays, instead of merely being bred within the single hives. I had to reply: No doubt you are right; but you will see with painful certainty—if not in thirty or forty, then certainty in forty to fifty years' time—that bee-keeping will thereby have been ruined.

These things must be considered. Everything is being mechanised and mineralised nowadays, but the fact is, the mineral world should only work in the way it does in Nature herself. You should not permeate the living Earth with something absolutely lifeless like the mineral, without including it in something else. It may not be possible to-morrow, but the day after to-morrow it will certainly be possible, quite as a matter of course.

Question: How should the insects be caught? Can they be used in the larval state?

Answer: You can use the larvae and the complete winged insect equally well. It may only involve a slight difference in the constellation. The proper constellation will move to some extent in the direction from Aquarius to Cancer as you pass from the winged insect to the larva. For the insect itself, the proper constellation will therefore be more towards Aquarius.

LECTURE SEVEN

KOBERWITZ,
June 15*th*, 1924.

MY DEAR FRIENDS,

In the remainder of the time at our disposal, I wish to say something about farm animals, orchards and vegetable gardening. We have not much time left; but in these branches of farming, too, we can have no fruitful starting-point unless we first bring about an insight into the *underlying* facts and conditions. We shall do this to-day, and pass on to-morrow to the more practical hints and applications.

To-day I must ask you to follow me in matters which lie yet a little farther afield from present-day points of view. Time was, indeed, when they were thoroughly familiar to the more instinctive insight of the farmer; to-day they are to all intents and purposes terra incognita. The entities occurring in Nature (minerals, plants, animals—we will leave man out for the moment) are frequently studied as though they stood there all alone.

Nowadays, one generally considers a single plant by itself. Then, from the single plant, one proceeds to consider a plant-species by itself; and other plant-species beside it. So it is all prettily pigeon-holed into species and genera, and all the rest that we are then supposed to know. Yet in Nature it is not so at all. In Nature—and, indeed, throughout the Universal being—all things are in mutual interaction; the one is always working on the other.

In our materialistic age, scientists only follow up the coarser effects of one upon the other—as for instance when one creature is eaten or digested by another, or when the dung of the animals comes on to the fields. Only these coarse interactions are traced. But in addition to these coarse interactions, finer ones, too, are constantly taking place—effects transmitted by finer forces and finer substances too—by warmth, by the chemical-ether principle that is for ever working in the atmosphere, and by the life-ether. . . .

We must take these finer interactions into account. Otherwise we shall make no progress in certain domains of our farm-work. Notably we must observe these more intimate relationships of Nature when we are dealing with the life, together on the farm, of plant and animal. Here again, we must not only consider those animals which are undoubtedly very near to us—like cattle, horses, sheep and so on. We must also observe with intelligence, let us say, the many-coloured world of insects, hovering around the plant-world during a certain season of the year. Moreover, we must learn to look with understanding at the birds.

Modern humanity has no idea how greatly *farming* and *forestry* are affected by the *expulsion*, owing to the modern conditions of life, of certain kinds of birds from certain districts. Light must be thrown upon these things once more by that *macrocosmic* method which Spiritual Science is pursuing—for we may truly call it macrocosmic. Here we can apply some of the ideas we have already let work upon us; we shall thus gain further insight.

Look at a *fruit-tree*—a pear-tree, apple-tree or plum-tree. Outwardly seen, to begin with, it is quite different from a herbaceous plant or cereal. Indeed, this would apply to any tree—it is quite different. But we must learn to perceive in what way the tree is different; otherwise we shall never understand the function of *fruit* in Nature's household (I am speaking now of such fruit as grows on trees).

Let us consider the tree. What is it in the household of Nature? If we look at it with understanding, we cannot include in the plant-nature of the tree any more than grows out of it in the thin stalks—in the green leaf-bearing stalks—and in the flowers and fruit. All this grows out of the tree, as the herbaceous plant grows out of the earth. The tree is really " earth " for that which grows upon its boughs and branches. It is the earth, grown up like a hillock; shaped—it is true—in a rather more living way than the earth out of which our herbaceous plants and cereals spring forth.

To understand the tree, we must say: There is the thick tree-trunk (and in a sense the boughs and branches still belong to this). Out of all this the real plant grows forth. Leaves, flowers and fruit grow out of this; they are the real plant—rooted in the trunk and branches of the tree, as the herbaceous plants and cereals are rooted in the Earth.

Here the question will at once arise: Is this " plant " which grows on the tree—and which is therefore describable as a parasitic growth, more or less—is it actually rooted? An actual root is not to be found in the tree. To understand the matter rightly, we must say: This plant which grows on the tree—unfolding up there its flowers and leaves and stems—has lost its roots. But a plant is not whole if it has no roots. It must have a root. Therefore we must ask ourselves: Where *is* the root of this plant?

The point is simply that the root is invisible to crude external observation. In this case we must not merely want to *see* a root—we must understand what a root is. A true comparison will help us forward here. Suppose I were to plant in the soil a whole number of herbaceous plants, very near together, so that their roots intertwined, and merged with one another—the one root winding round the other, until it all became a regular mush of roots, merging one into another. As you can well imagine, such a complex of roots would not allow itself to remain a mere tangle; it would grow organised into a single entity. Down there in the soil the saps and fluids would flow into one another. There would be an organised root-complex—roots flowing into one another. We could not distinguish where the several roots began or ended. A common root-being would arise for these plants (Diagram 15).

So it would be. No such thing need exist in reality, but this illustration will enable us to understand. Here is the soil of the earth: here I insert all my plants. Down there, all the roots coalesce, until they form a regular surface—a continuous root-stratum. Once more, you would not know where the one root begins and the other ends.

Now the very thing I have here sketched as an hypothesis is actually present in the tree. The plant which grows on the tree has lost its root. Relatively speaking, it is even separated from its root—only it is united with it, as it were, in a more ethereal way. What I have hypothetically sketched on the board is actually there in the tree, as the cambium layer—the *cambium*. That is how we must regard the roots of these plants that grow out of the tree: they are replaced by the cambium. Although the cambium does not look like roots, it is the living, growing layer, constantly forming new cells, so that the plant-life of the tree grows out of it, just as the life of a herbaceous plant grows up above out of the root below.

Here, then, is the tree with its cambium layer, the growing formative layer, which is able to create plant-cells. (The other layers in the tree would not be able to create fresh cells). Now you can thoroughly see the point. In the tree with its cambium or formative layer, the earth-realm itself is actually bulged out; it has grown outward into the airy regions. And having thus grown outward into the air, it needs more inwardness, more intensity of life, than the earth otherwise has, *i.e.* than it has where the ordinary root is in it. Now we begin to understand the tree. In the first place, we understand it as a strange entity whose function is to separate the plants that grow upon it—stem, blossom and fruit—from their roots, uniting them only through the Spirit, that is, through the ethereal. We must learn to look with macrocosmic intelligence into the mysteries of growth. But it goes still further. For I now beg you observe: What happens through the fact that a tree comes into being? It is as follows:—

That which encompasses the tree has a different plant-nature in the air and outer warmth than that which grows in air and warmth immediately on the soil, unfolding the herbaceous plant that springs out of the earth directly (Diagram 16). Once more, it is a different plant-world. For it is far more intimately related to the surrounding *astrality*. Down here, the astrality in air and warmth is expelled, so that the air and warmth may become mineral for the sake of man and animal. Look at a plant growing directly out of the soil. True, it is hovered-around, enshrouded in an astral cloud. Up there, however, round about the tree, the astrality is far denser. Once more, it is far denser. Our *trees* are gatherings of astral substance; quite clearly, they are *gatherers of astral substance*.

In this realm it is easiest of all for one to attain to a certain higher development. If you make the necessary effort, you can easily become esoteric in these spheres. I do not say *clairvoyant*, but you can easily become clair-sentient with respect to the sense of smell, especially if you acquire a certain sensitiveness to the diverse aromas that proceed from plants growing on the soil, and on the other hand from fruit-tree plantations—even if only in the blossoming stage—and from the woods and forests! Then you will feel the difference between a plant-atmosphere poor in astrality, such as you can smell among the herbaceous plants growing on the earth, and a plant-world rich in astrality such as you have in your nostrils when you sniff what is so beautifully wafted from the tree-tops.

Accustom yourself to specialise your sense of smell—to distinguish, to differentiate, to individualise, as between the scent of earthly plants and the scent of trees. Then, in the former case you will become clair-sentient to a thinner astrality, and in the latter case to a denser astrality. You see, the farmer can easily become clair-sentient. Only in recent times he has made less use of this than in the time of the old clairvoyance. The countryman, as I said, can become clair-sentient with regard to the sense of smell.

Let us observe where this will lead us. We must now ask: What of the polar opposite, the counterpart of that richer astrality which the plant—parasitically growing on the tree—brings about in the neighbourhood of the tree? In other words, what happens by means of the cambium? What does the cambium itself do?

Far, far around, the tree makes the spiritual atmosphere inherently richer in astrality. What happens, then, when the herbaceous life grows out of the tree up yonder? The tree has a certain inner vitality or ethericity; it has a certain intensity of life. Now the cambium damps down this life a little more, so that it becomes slightly more mineral. While, up above, a *rich astrality* arises all around the tree, the cambium works in such a way that, there within, the ethericity is poorer.

Within the tree arises poverty of ether as compared to the plant. Once more, here within, it will be somewhat poorer in ether. And as, through the cambium, a relative *poverty of ether* is engendered in the tree, the root in its turn will be influenced. The roots of the tree become mineral—far more so than the roots of herbaceous plants. And the root, being more mineral, deprives the earthly soil —observe, we still remain within the realms of life—of some of its ethericity. This makes the earthly *soil* rather *more dead* in the environment of the tree than it would be in the environment of a herbaceous plant.

All this you must clearly envisage. Now whatever arises in this way will always involve something of deep significance in the household of Nature as a whole. Let us then enquire: what is the inner significance, for Nature, of the astral richness in the tree's environment above, and the etheric poverty in the realm of the tree-roots? We only need look about us, and we can find how these things work themselves out in Nature's household. The fully developed *insect*, in effect, lives and moves by virtue of this rich astrality which is wafted through the tree-tops.

Take, on the other hand, what becomes poorer in ether, down below in the soil. (This poverty of ether extends, of course, throughout the tree, for the Spiritual always works through the *whole*, as I explained yesterday when speaking of human Karma). That which is poorer in ether, down below, works through the *larvae*. Thus, if the earth had no trees, there would be no insects on the earth. The trees make it possible for the insects to be. The insects fluttering around the parts of the tree which are above the earth—fluttering around the woods and forests as a whole—they have their very life through the existence of the woods. Their larvae, too, live by the very existence of the woods.

Here you have a further indication of the inner relationship between the root-nature and the sub-terrestrial animal world. From the tree we can best learn what I have now explained; here it becomes most evident. But the fact is: What becomes very evident in the tree is present in a more delicate way throughout the whole plant-world. In every plant there is a certain tendency to become tree-like. In every plant, the root with its environment strives to let go the ether; while that which grows upward tends to draw in the astral more densely. The tree-becoming tendency is there is every plant.

Hence, too, in every plant the same relationship to the insect-world emerges, which I described for the special case of the tree. But that is not all. This relation to the insect-world expands into a relation to the whole animal kingdom. Take, for example, the insect larvae: truly, they only live upon the earth by virtue of the tree-roots being there. However, in times gone by, such larvae have also evolved into other kinds of animals, similar to them, but undergoing the whole of their animal life in a more or less larval condition. These creatures then emancipate themselves, so to speak, from the tree-root-nature, and live more near to the rest of the root-world—that is, they become associated with the root-nature of herbaceous plants.

A wonderful fact emerges here: Certain of these sub-terrestrial creatures (which, it is true, are already somewhat removed from the larval nature) develop the faculty to regulate the ethereal vitality within the soil whenever it becomes too great. If the soil is tending to become too strongly living—if ever its livingness grows rampant—these sub-terranean animals see to it that the over-intense vitality is released. Thus they become wonderful *regulators*, safety-valves for the vitality inside the Earth. These golden creatures—for they are of the greatest value to the earth—are none other than the *earth-worms*.

Study the earth-worm—how it lives together with the soil. These worms are wonderful creatures: they leave to the earth precisely as much *ethericity* as it needs for plant-growth. There under the earth you have the earth-worms and similar creatures—distantly reminiscent of the larva. Indeed, in certain soils—which you can easily tell—we ought to take special care to allow for the due breeding of earth-worms. We should soon see how beneficially such a control of the animal world beneath the earth would react on the vegetation, and thus in turn upon the animal world in general, of which we shall speak in a moment.

Now there is again a distant similarity between certain animals and the fully evolved, *i.e.* the winged, insect-world. These animals are the *birds.* In course of evolution a wonderful thing has taken place as between the insects and the birds. I will describe it in a picture. The insects said, one day: We do not feel quite strong enough to work the astrality which sparkles and sprays around the trees. We therefore, for our part, will use the treeing tendency of other plants; there we will flutter about, and to you birds we will leave the astrality that surrounds the trees. So there came about a regular division of labour between the *bird-world* and the *butterfly-world,* and now the two together work most wonderfully.

These winged creatures, each and all, provide for a proper distribution of *astrality,* wherever it is needed on the surface of the Earth or in the air. Remove these winged creatures, and the astrality would fail of its true service; and you would soon detect it in a kind of stunting of the vegetation. For the two things belong together: the winged animals, and that which grows out of the Earth into the air. Fundamentally, the one is unthinkable without the other. Hence the farmer should also be careful to let the insects and birds flutter around in the right way. The farmer himself should have some understanding of the care of *birds* and *insects.* For in great Nature—again and again I must say it—everything, everything is connected.

These things are most important for a true insight: therefore let us place them before our souls most clearly. Through the flying world of insects, we may say, the right astralisation is brought about in the air. Now this astralisation of the air is always in mutual relation to the woods or forests, which guide the astrality in the right way—just as the blood in our body is guided by certain forces. What the wood does—not only for its immediate vicinity but far and away around it (for these things work over wide areas)—what the wood does in this direction has to be done by quite other things in unwooded districts. This we should learn to understand. The growth of the soil is subject to quite other laws in districts where forest, field and meadow alternate, than in wide, unwooded stretches of country.

There are districts of the Earth where we can tell at a glance that they became rich in forests long before man did anything—for in certain matters Nature is wiser than man, even to this day. And we may well assume, if there is forest by Nature in a given district, it has its good use for the surrounding farmlands—for the herbaceous and graminaceous vegetation. We should have sufficient insight, on no account to exterminate the forest in such districts, but to preserve it well. Moreover, the Earth by and by changes, through manifold cosmic and climatic influences.

Therefore we should have the heart—when we see that the vegetation is becoming stunted, not merely to make experiments for the fields or on the fields alone, but to increase the wooded areas a little. Or if we notice that the plants are growing rampant and have not enough seeding-force, then we should set to work and make some clearings in the forest—take certain surfaces of wooded land away. In districts which are pre-destined to be wooded, the *regulation of woods* and *forests* is an essential part of agriculture, and should indeed be thought of from the spiritual side. It is of a far-reaching significance.

Moreover, we may say: the world of worms, and larvae too, is related to the *limestone*—that is, to the mineral nature of the earth; while the world of insects and birds—all that flutters and flies—stands in relation to the astral. That which is there under the surface of the earth—the world of worms and larvae—is related to the mineral, especially the chalky, limestone nature, whereby the ethereal is duly conducted away, as I told you a few days ago from another standpoint. This is the task of the limestone—and it fulfils its task in mutual interaction with the larva- and insect-world.

Thus you will see, as we begin to specialise what I have given, ever new things will dawn on us—things which were undoubtedly recognised with true feeling in the old time of instinctive clairvoyance. (I should not trust myself to expound them with equal certainty.) The old instincts have been lost. Intellect has lost all the old instincts—nay, has exterminated them. That is the trouble with materialism—men have become so intellectual, so clever. When they were less intellectual, though they were not so clever, they were far wiser; out of their feeling they knew how to treat things, even as we must learn to do once more, for in a conscious way we must learn once more to approach the Wisdom that prevails in all things. We shall learn it by something which is not clever at all, namely, by Spiritual Science. Spiritual Science is not clever: it strives rather for Wisdom.

Nor can we rest content with the abstract repetition of words: " Man consists of physical body, etheric body," etc., etc., which one can learn off by heart like any cookery-book. The point is for us to introduce the knowledge of these things in all domains—to see it inherent everywhere. Then we are presently guided to distinguish how things are in Nature, especially if we become clairvoyant in the way I explained. Then we discover that the *bird world* becomes harmful if it has not the " needle-woods " or *coniferous* forests beside it, to transform what it brings about into good use and benefit. Thereupon our vision is still further sharpened, and a fresh relationship emerges. When we have recognised this peculiar relation of the birds to the coniferous forests, then we perceive another kinship. It emerges clearly. To begin with, it is a fine and intimate kinship—fine as are those which I have mentioned now. But it can readily be changed into a stronger, more robust relationship.

I mean the inner kinship of the *mammals* to all that does not become tree and yet does not remain as a small plant—in other words, to the *shrubs* and *bushes*—the hazel-nut, for instance. To improve our stock of mammals in a farm or in a farming district, we shall often do well to plant in the landscape bushes or shrub-like growths. By their mere presence they have a beneficial effect. All things in Nature are in mutual interaction, once again. But we can go farther. The animals are not so foolish as men are; they very quickly " tumble to it " that there is this kinship. See how they love the shrubs and bushes. This love is absolutely inborn in them, and so they like to get at the shrubs to eat them. They soon begin to take what they need, which has a wonderfully regulating effect on their remaining fodder.

Moreover, when we trace these intimate relationships in Nature, we gain a new insight into the essence of what is harmful. For just as the coniferous forests are intimately related to the birds and the bushes to the mammals, so again all that is *mushroom*—or fungus-like—has an intimate relation to the lower animal world—to the bacteria and such-like creatures, and notably the harmful parasites. The harmful parasites go together with the mushroom or fungus-nature; indeed they develop wherever the fungus-nature appears scattered and dispersed.

Thus there arise the well-known plant-diseases and harmful growths on a coarser and larger scale. If now we have not only woods but meadows in the neighbourhood of the farm, these meadows will be very useful, inasmuch as they provide good soil for mushrooms and toadstools; and we should see to it that the soil of the meadow is well-planted with such growths. If there is near the farm a meadow rich in mushrooms—it need not even be very large—the mushrooms, being akin to the bacteria and other parasitic creatures, will keep them away from the rest. For the mushrooms and toadstools, more than the other plants, tend to hold together with these creatures. In addition to the methods I have indicated for the destruction of these pests, it is possible on a larger scale to keep the harmful microscopic creatures away from the farm by a proper distribution of meadows.

So we must look for a due distribution of wood and forest, orchard and shrubbery, and meadow-lands with their natural growth of mushrooms. This is the very essence of good farming, and we shall attain far more by such means, even if we reduce to some extent the surface available for tillage.

It is no true economy to exploit the surface of the earth to such an extent as to rid ourselves of all the things I have here mentioned in the hope of increasing our crops. Your large plantations will become worse in quality, and this will more than outweigh the extra amount you gain by increasing your tilled acreage at the cost of these other things. You cannot truly engage in a pursuit so intimately connected with Nature as farming is, unless you have insight into these mutual relationships of Nature's husbandry.

The time has come for us to bring home to ourselves those wider aspects which will reveal, quite generally speaking, the relation of plant to animal-nature, and *vice versa*, of animal to plant-nature. What is an animal? What is the world of plants? (For the world of plants we must speak rather of a totality—the plant-world as a whole.) Once more, what is an animal, and what is the world of plants? We must discover what the essential relation is; only so shall we understand how to feed our animals. We shall not feed them properly unless we see the true relationship of plant and animal. What *are* the animals? Well may you look at their outer forms! You can dissect them, if you will, till you get down to the skeleton, in the forms of which you may well take delight; you may even study them in the way I have described. Then you may study the musculature, the nerves and so forth.

All this, however, will not lead you to perceive what the animals really are in the whole household of Nature. You will only perceive it if you observe what it is in the environment to which the animal is directly and intimately related. What the animal receives from its environment and assimilates directly in its nerves-and-senses system and in a portion of its breathing system, is in effect all that which passes first through *air* and *warmth*. Essentially, in its own proper being, the animal is a direct assimilator of *air* and *warmth*—through the nerves-and-senses system.

Diagrammatically, we can draw the animal in this way: In all that is there in its periphery, in its environment—in the nerves-and-senses system and in a portion of the breathing system—the animal is itself. In its own essence, it is a creature that lives directly in the air and warmth. It has an absolutely direct relation to the air and warmth (Diagram 17).

Notably out of the warmth its bony system is formed—where the Moon- and Sun-influences are especially transmitted through the warmth. Out of the air, its muscular system is formed. Here again, the forces of Sun and Moon are working through the air. But the animal cannot relate itself thus directly to the *earthy* and *watery* elements. It cannot assimilate water and earth thus directly. It must indeed receive the earth and water into its inward parts; it must therefore have the digestive tract, passing inward from outside. With all that it has become through the warmth and air, it then assimilates the water and the earth inside it—by means of its metabolic and a portion of its breathing system.

The breathing system passes over into the metabolic system. With a portion of the breathing and a portion of the metabolic system, the animal assimilates " earth " and " water." In effect, before it can assimilate earth and water, the animal itself must be there *by virtue of* the air and warmth. That is how the animal lives in the domain of earth and water. (The assimilation-process is of course, as I have often indicated, an assimilation more of forces than of substances).

Now let us ask, in face of the above, what is a plant? The answer is: the plant has an immediate relation to earth and water, just as the animal has to air and warmth. The plant—also through a kind of breathing and through something remotely akin to the sense-system—absorbs into itself directly all that is earth and water; just as the animal absorbs the air and warmth. The plant lives directly with the earth and water.

Now you may say: Having recognised that the plant lives directly with earth and water, just as the animal does with air and warmth, may we not also conclude that the plant assimilates the air and the warmth internally, even as the animal assimilates the earth and water? No, it is not so. To find the spiritual truths, we cannot merely conclude by analogy from what we know. The fact is this: Whereas the animal consumes the earthy and watery material and assimilates them internally, the plant does not consume but, on the contrary, secretes—gives off—the air and warmth, which it experiences in conjunction with the earthy soil. Air and warmth, therefore, do not go in—at least, they do not go in at all far. On the contrary they go out; instead of being consumed by the plant, they are given off, excreted, and this excretion-process is the important point.

Organically speaking, the plant is in all respects an inverse of the animal—a true inverse. The excretion of air and warmth has for the plant the same importance as the consumption of food for the animal. In the same sense in which the animal lives by absorption of food, the plant lives by excretion of air and warmth. This, I would say, is the virginal quality of the plant. By nature, it does not want to consume things greedily for itself, but, on the contrary, it gives away what the animal takes from the world, and lives thereby. Thus the plant gives, and lives by giving.

Observe this give and take, and you perceive once more what played so great a part in the old instinctive knowledge of these things. The saying I have here derived from anthroposophical study: " The plant in the household of Nature gives, and the animal takes," was universal in an old instinctive and clairvoyant insight into Nature. In human beings who were sensitive to these things, some of this insight survived into later times.

In Goethe you will often find this saying: Everything in Nature lives by give and take. Look through Goethe's works and you will soon find it. He did not fully understand it any longer, but he revived it from old usage and tradition; he felt that this proverb describes something very true in Nature. Those who came after him no longer understood it. To this day they do not understand what Goethe meant when he spoke of "give and take." Even in relation to the breathing process—its interplay with the metabolism—Goethe speaks of "give and take." Clearly-unclearly, he uses this word.

Thus we have seen that forest and orchard, shrubbery and bush are in a certain way regulators to give the right form and development to the growth of plants over the earth's surface. Meanwhile beneath the earth the lower animals—larvae and worm-like creatures and the like, in their unison with limestone—act as a regulator likewise.

So must we regard the relation of tilled fields, orchards and cattle-breeding in our farming work. In the remaining hour that is still at our disposal, we shall indicate the practical applications, enough for the good Experimental Circle to work out and develop.

LECTURE EIGHT

KOBERWITZ,
16th June, 1924.

MY DEAR FRIENDS,

This is our last lecture, though we may still be able to supplement it a little in the discussions, according to your needs. As far as possible in the short time, I want to add a few more explanations to complete what I have said, and to give a few more practical hints. These practical matters are, however, extremely difficult to clothe in general formulae or the like. They, most of all, are subject to individualisation—to a kind of personal treatment. To-day especially, we shall therefore have to acquire the necessary spiritual-scientific insight to begin with, for this alone will enable you to act with individual intelligence in the several measures you have to take.

Think how little insight there is nowadays in this most important question: the feeding of farm animals. Such a state of affairs cannot be much improved by however many detailed instructions for feeding. But I am convinced it will be much improved when our agricultural training tends more to the development of true insight on the fundamental question: What is the essence of the feeding process? To-day I would like to contribute a little to this end.

As I have already told you, the significance of nutrition for the animal, and for man too, is to this day thoroughly misunderstood. The coarse idea that the foodstuffs are received from outside and then deposited in the organism, is altogether wrong. That is what they imagine nowadays, more or less. True, they conceive all kinds of transformations in the process, and yet, fundamentally speaking, that is how they think. In a crude way they imagine, somewhere inside there are the foodstuffs. The animal absorbs the food—deposits inside it whatever it can use, and excretes what it has no use for. Accordingly, they argue, we must provide for such and such essential constituents. We must see to it that the creature is not over-burdened with stuff. We must see to it that the food it gets is as nutritive as possible, so that it can use a relatively large proportion of what is contained therein.

True, they also distinguish between substances nutritive in the narrower sense of the term, and those which—as they say—assist the combustion-process in the body. (The materialists are fond of making such distinctions also). On this distinction they found all manner of theories which they then apply in practice, though as you know, the upshot always is that some of it works and some of it decidedly does not—or it only seems to work for a limited time, and is then modified by this or that. . . .

And how should we expect it to be otherwise? They talk of combustion-processes inside the body. In reality there is not a single combustion-process in the body. The combination of any substance with oxygen *inside the body* has quite another significance than that of a combustion-process. Combustion is a process in mineral or lifeless Nature. Quite apart from the fact that a living organism is essentially different from a crystal of quartz, what is commonly called combustion in the body is not the dead combustion-process which takes place in the outer world, but is something altogether living, nay, sentient.

Precisely by expressing themselves in this way, and thus leading people's thought in a fixed direction, scientists bring about widespread confusion in practical life. The man who first speaks of " combustion inside the body " is only speaking loosely—in a slipshod way, if you will. If he has the true facts in mind, his speaking loosely will do no harm, provided he still acts correctly, out of true instincts or tradition. After a time, however, the same loosely worded phrase gets taken hold of by the disease of " Psychopathia professoralis," as I have often called it. They—the professors—transform, what at first was only a slight slipshod way of talking into a brilliant theory —I really mean it, brilliant. And when people begin to act according to these theories, they no longer hit off the reality in the very least. The things they then talk of are altogether different from what actually occurs when you have animals to look after. It is a characteristic phenomenon of to-day. They set to work and do something utterly different—something that does not fit in at all with what is actually taking place in Nature. In this domain especially, we should take pains to observe what the point is.

Let us remember the outcome of our last lecture. The plant, as we saw, has a physical body and an ether-body, while up above it is hovered-around, more or less, by a kind of astral cloud. The plant itself does not reach up to the astral, but the astral—so to speak—hovers around it. Wherever it enters into definite connection with the astral (as happens in the fruit-formation), something available as foodstuff is produced—that is to say, something which will support the astral in the animal and human body.

If you see into the process, you will readily observe in any plant or other entity, whether or no it is fit to support some process in the animal organism. But we should also understand the opposite pole. This is a most important point; I have already touched upon it, but now that we wish to create the foundations for an understanding of the feeding-process, we must bring it out once more with special emphasis. As we are now concerned with the feeding problem, let us begin with the animal.

In the animal there is no such sharply outlined threefolding of the organism as there is in man. True, in the animal also, the nerves-and-senses organism and the organism of metabolism, and the limbs are well marked—sharply divided one from the other; but the middle, rhythmic organism more or less melts away—at least, in many animals it does so. Something that still comes from the sense-organism passes into the rhythmic; likewise, something that comes from the metabolic organism.

We should describe the animal rather differently from man. For man, we speak quite exactly when we describe this threefold nature of the body; for the animal, however, we should rather speak as follows: There is the nerves-and-senses organisation, mainly localised in the head. There is the organisation of metabolism and the limbs—organised in the posterior parts and in the limbs generally, yet also permeating the whole body. And in the middle of the creature the metabolism becomes rhythmic—more rhythmic than in man; while on the other hand the nerves-and-senses system also becomes more rhythmic, and the two melt into one another. In other words, the rhythmic part of the animal does not come into being so independently as in man; it is a more indistinct sounding-into-one-another of the two outermost poles (Diagram 18). Hence, for the animal we should really speak of a *two-foldness* of the organism—such, however, that the two members meet and mingle in the middle. That is how the animal organisation arises.

Now all that is present as substances in the head-organisation, is composed of earthly matter. (So it is in man, too, but let us confine ourselves to the animal for the moment). Whatever matter there is in the head is earthly matter. Already in the embryo-life, earthly matter is guided into the head-organisation. The whole embryonic organisation is so arranged that the head receives its materials from the Earth. There, then, we have earthly substance.

On the other hand, all that we have as substantiality in the organisation of metabolism and the limbs—permeating our intestines, limbs, muscles, bones, etc.—comes not from the Earth at all. It is cosmic substantiality. It comes from that which is absorbed out of the air and warmth above the Earth. This is important. You must not regard a claw or a hoof as though it were formed by the physical matter which the animal eats somehow finding its way into the hoof and being there deposited. That is not true at all. In actual fact, cosmic matter is absorbed through the senses and the breathing. What the animal eats is merely for the purpose of developing its inner forces of movement, so that the cosmic principles may be driven right down into the metabolic and limb-system—into the claw or hoof, for instance. Throughout these parts, it is cosmic substantiality.

Precisely the opposite is true of the *forces*. In the head—inasmuch as the senses are chiefly stationed there, and the senses perceive out of the Cosmos—in the head we have cosmic forces; while in the system of metabolism and limbs we have to do with earthly forces—cosmic substances and earthly forces. (As to the latter, you need only remember how we walk; we are constantly placing ourselves into the field of earthly gravity, and in like manner, all that we do with our limbs is bound up with the earthly).

This is by no means a matter of indifference, in practice. Suppose you are using the cow as a beast of labour. It needs its limbs for the work. Or if you use an ox as a labouring beast—it is important to feed the animal so that it gets as much as possible of cosmic substantiality. Moreover, the food which will pass through the stomach must be suitably chosen and arranged, so as to develop copious forces—forces sufficient to guide the cosmic substantiality into the limbs and bones and muscles, everywhere.

Likewise we need to be aware: whatever substances are required for the head itself—these must be got from the actual fodder. The foodstuffs—assimilated, passed through the stomach—must be guided into the head. It is the head, not the big toe, which depends on the stomach in this respect! Moreover, the head can only assimilate this nourishment which it received from the body, if it is able at the same time to get the necessary forces from the Cosmos. Therefore we should not merely shut our animals in dark stables, where the cosmic forces cannot flow towards them. We should lead them out over the pastures. Altogether, we should give them the opportunity to come into relation with the surrounding world by sense-perception too.

Think of an animal standing in the dark, dull stable, and receiving—measured out into its manger—what the wisdom of man provides. Such an animal, getting no change in this respect, and it *can* only get the proper change in the open air—how different it will be from one that is able to make use of its senses, its organ of smell, for instance, seeking its food for itself in the open air; following its sense of smell, following the cosmic forces through its sense of smell, going after the food, choosing for itself, unfolding all its activity in this finding and taking of the food.

Such things are inherited. The animal you merely place at the manger will not reveal at once that it has no cosmic forces; for it still inherits them. But it will presently beget descendants which have the cosmic forces in them no longer. In such a case, it is from the head that the animal first becomes weak. It can no longer feed the body because it is unable to absorb the cosmic substances, which, once again, are needed in the body as a whole.

These things will show you how futile it is merely to give general instructions: " Feed thus and thus, in this case and in that." We must first gain an idea: what is the value of such and such methods of feeding for the whole essence of the animal's organisation?

Now we can go further. What is contained in the head? Earthly substantiality. Cut out this noblest organ of the animal—the brain—there you have so much earthly substance. In man, too, in the brain you have earthly substance. Only the forces are cosmic; the substance is earthly. What then is the function of the brain? It serves as an underlying basis for the Ego. The animal has not yet the Ego. Let us hold fast to this idea: The brain serves as an underlying basis for the Ego, but the animal has not yet an Ego. Therefore the animal's brain is only on the way to Ego-formation. In man it goes on and on—to the full forming of the Ego.

How then has the brain of the animal come into being? Take the whole organic process—all that is going on in there. That which eventually emerges as earthly matter in the brain has actually been excreted; it is excretion—excretion from the organic process. Earthly matter is here excreted to serve as a basis for the Ego. Now on the basis of this process in the metabolic and limbs-system—beginning with the consumption of the food and going on through the whole distributive activity of the digestion—a certain quantity of earthly matter is capable of being led into the head and brain. A certain quantity of earthly substance goes through the whole path, and is at last literally deposited—excreted, separated out—in the brain. But it is not only in the brain that the substance of the foodstuffs is deposited. Whatever is no longer capable of assimilation is deposited already on the way, in the intestines.

Here you encounter a relationship which you will think most paradoxical, even absurd at first sight, and yet you cannot overlook it if you wish to understand the animal organisation—and the human too, for that matter. What is this brainy mass? It is simply an intestinal mass, carried to the very end. The premature brain-deposit passes out through the intestines. As to its processes, the content of the intestines is decidedly akin to the brain-content. To speak grotesquely, I would say: That which spreads out through the brain is a highly advanced heap of manure! Grotesque as it may be, objectively speaking this is the truth. It is none other than the dung, which is transmuted—through its peculiar organic process—into the noble matter of the brain, there to become the basis for Ego-development.

In man, as much as possible of the belly-manure is transformed into brain-manure, for man as you know carries his Ego down on to the Earth; in the animal, less. Therefore, in the animal, more remains behind in the belly-manure—and this is what we use for manuring. In animal manure, *more* Ego potentially remains. Just *because* the animal itself does not reach up to the Ego, more Ego remains there potentially. Hence, animal and human manure are altogether different things. Animal manure still contains the Ego-potentiality.

Picture to yourselves how we manure the plant. We bring the manure from outside to the plant root. That is to say, we bring *Ego* to the root of the plant. Let us draw the plant in its entirety (Diagram 19). Down here you have the root; up there, the unfolding leaves and blossoms. There, through the intercourse with air, astrality unfolds—the astral principle is added—whereas down here, through intercourse with the manure, the Ego-potentiality of the plant develops.

Truly, the farm is a living organism. Above, in the air, it evolves its astrality. Fruit-tree and forest by their very presence develop this astrality. And now when the animals feed on what is there above the Earth, they in their turn develop the real Ego-forces. These they give off in the dung, and the same Ego-forces will cause the plant in its turn to grow forth from the root in the direction of the force of gravity. Truly a wonderful interplay, but we must understand it stage by stage, progressively, increasingly.

Inasmuch as these things are so, your farm is in truth a kind of individuality, and you will gain the insight that you ought to keep your animals as much as possible *within* this mutual interplay—and your plants too. Thus, in a sense, you mar the working of Nature when you take your manure not from your own farm animals, but get rid of the animals and order the manure-content from Chile. Then you are playing fast and loose with things—neglecting the fact that this is a perfect and self-contained cycle, which ought to be maintained, complete in itself. Needless to say, we must *arrange* things so; we must have enough and the right kind of animals, so as to get enough manure and the right kind for our farm. Or again, we must take care to plant what the animals which we desire to have will like to eat instinctively—what they will seek out for themselves. Naturally, here our experiments grow complicated—they become individual, in fact.

Hence, as I said, we must first indicate general guiding lines for individual treatment. Much will remain to be tried out. Then useful rules of conduct will emerge; but all of these will proceed from the one guiding line: to make the farm, as far as possible, so self-contained that it is able to sustain itself. As far as possible—not quite! Why not? The concrete study of Spiritual Science will never make you a fanatic. In outer life, within our present economic order, it cannot be fully attained. Nevertheless, you should try to attain it as far as possible.

We can now find the concrete, specific relations of the animal organism to the plant—that is, to the organism of the fodder. Let us first see it as a whole. Observe the root, which develops as a rule inside the earth. There the manure permeates it, as we have seen, with a nascent Ego-force—an Ego-force in process of becoming. Through the whole way it lives in the Earth, the root absorbs this nascent Ego-force. The root is assisted in absorbing this Ego-force if it can find the proper quantity of salt in the Earth. Here then we have the root. Simply on the basis of the thoughts we have already placed before us, we can now recognise it as that foodstuff which, if it comes into the human organism, will most easily find its way, in the digestive process, to the head.

We shall therefore provide root-nourishment if we must assume that substance—material substance—is needed for the head, so that the cosmic forces working plastically through the head may find the proper stuff to work upon. What will it remind you of when this is said: " I must give roots as fodder to an animal which needs to carry material substance into its head, so that it may have a live and mobile sense-relationship, *i.e.* a cosmic relationship, to its cosmic environment." Will you not immediately think of the calf and the carrot? When the calf eats the carrot, this process is fulfilled.

You see, the moment you express such a piece of knowledge—if you are actually aware what a farm looks like, what it is like in practice, your thoughts will turn at once to what is actually done. You need only know that this is the real mutual process.

Let us proceed. Now that the material substance has been conveyed into the head—now that we have served the calf with the carrot—the reverse process must be able to take place. The head must be able to work with will-activity, creating forces in the organism, so that these forces in their turn can work right down into the body. The carrot-dung must not be merely deposited in the head. *From* what is there deposited—from what is there in process of disintegration—force-radiations must pass into the body. Therefore you need a second foodstuff. Having now served this member of the body, you need a second foodstuff which in its turn will enable the head to fulfil its proper function by the remainder of the body.

Suppose, then, I have given carrot-fodder. Now I want the body to be properly permeated by the forces that are able to evolve out of the head. Now I need something in Nature that has a ray-like, radiating form, or that gathers up the ray-like nature in a concentrated " tabloid " form, so to speak. What shall I use, then, as a second foodstuff? Once more, I shall add to the carrot something that tends to ray out in the plant, and afterwards gathers-in its ray-like force in concentration. So my attention is directed to linseed or the like. Such is the fodder you should give young cattle. Carrots and linseed, or something that will go together on the same principle—say, for instance, carrots and fresh hay. These will work through and through the animal—mastering its inner processes—setting it well on the way of its development.

Thus, for young cattle, we shall always try to provide fodder such as will stimulate the Ego-forces on the one hand, and on the other hand assist what passes downward from above—the astral radiations which are needed to fill the body through. Assistance of the latter kind is rendered especially by long and thin-stalked plants, left simply to their own development—that is to say, long grass, etc., that has grown into hay—whatever is long and thin-stalked and goes to hay (Diagram 20). In agriculture we must always learn to look at the things themselves, and of each thing we must learn what happens to it when it passes, either from the animal into the soil, or from the plant into the animal.

Let us pursue the matter further. Suppose you wish the animal to become strong precisely in the middle region, where the head-organisation—that of nerves-and-senses—develops more towards the breathing, and on the other hand the metabolic organisation also tends towards the rhythmic life, and the two poles interpenetrate. What animals do you wish to become strong in this region? The milk-giving creatures—they must grow strong in this middle part. For in the production of *milk* precisely this requirement is fulfilled.

What must you care for in this case? You must see that the right co-operation is there between the stream that passes backwards from the head—which is mainly a streaming of *forces*—and the stream that passes forward from behind, which is mainly a streaming of *substance*. If this co-operation is taking place, so that the streaming from behind is thoroughly worked through by the forces that flow from the fore-parts backward, good and copious milk will be the outcome. For the good milk contains what has been specially developed in the metabolic process. It is a metabolic preparation, which, though it has not yet passed through the sexual system, has become as nearly as possible akin—in the digestive process itself—to the sexual digestive process. Milk is a transformed sexual gland secretion. A substance which is on the way to become sexual secretion is met by the head-forces working into it and so transforming it. You can see right into this process.

If now we wish the processes to form themselves in this way, we must look around for foodstuffs working less towards the head than the roots, which latter have absorbed the Ego-force. At the same time, since it has to remain akin to the sexual force, we must not have too much of the astral in it—not too much of what tends towards blossom and fruit. For a good milk-production we must therefore look to what is there between the flower and the root—that is, to the green foliage: all that unfolds in leaf and vegetable foliage (Diagram 21).

If we want to stimulate the development of milk, in an animal whose milk-production we have reason to believe could be increased, we shall certainly attain the desired end if we proceed as follows. Assume I am feeding a milch cow—according to the prevailing conditions—with vegetable leaves or foliage or the like. Now I want to increase the milk production. I say to myself, it surely can be increased. What shall I do? I shall use plants which draw the fruiting process—the process that takes place in flower and fertilisation—down into the foliage, into the leafing process. This applies for instance to the pod-bearing or leguminous plants—notably the various kinds of clover. In the clover-substance, manifold elements of a fruit-like quality develop just life leaf and foliage.

Treat the cow in this way and you will not see much result in the cow herself, but when she calves (for the fodder-reforms you introduce along these lines generally take a generation to work themselves out), when the cow calves, the calf will become a good milch-cow.

One thing especially you must observe in all these matters. As a rule, when the traditions of old instinctive wisdom vanished from this sphere, a few things were maintained—just as our doctors have maintained a few of the old remedies. Though they no longer know why, they have kept them on, simply because they always find them helpful. Likewise in farming, certain things are known out of old tradition. People do not know why, but they continue to use them, and for the rest, they make experiments and tests. Thus they try to indicate the quantities that should be given for fattening cattle, milch cattle and the like. But the whole thing turns out as it usually does when men begin to experiment at random—especially when their experimenting is left to mere chance.

Think what happens, for example, if ever you have a sore throat at a place where you are among many people. Everyone who is fond of you will offer you some remedy. Within half an hour you have a whole chemist's shop! If you really took all these remedies the one would cancel the other out, and the only sure thing is that you would suffer indigestion, while your sore throat would be no better. The simple measures that ought to be taken are thus transformed into great complication.

So it is when you begin experimenting with all kinds of fodder. You begin to use something. In a certain direction it goes well, in another it does not. Now you add a second fodder to it, and so you go on, and the result is a whole number of standard fodders, each of which has its significance for young cattle or fattening cattle as the case may be, but it all becomes very complicated, and to-day no one can see the wood for the trees. They have no longer any comprehensive vision of the relationships of forces which are involved. Or again, the effect of the one thing is such as to cancel the other out.

This is happening very widely, especially among those who have acquired a little learning by their academic studies, and thereupon go out and try to farm. Then they look up their text-books, or they remember what they learned: " Young cattle should be fed so and so, cattle you wish to fatten should be fed in that way," and so on. So they will look it all up. But the results will not be very great, for it may easily happen that what you look up in the text-book will clash with what you are already giving of your own accord.

You can only proceed rationally by taking your start from a way of thought such as I have now indicated, for this will very largely simplify the animal's food, and you will gain a comprehensive view of what you are doing. For instance, you can see quite clearly and straightforwardly that carrots and linseed together will work in this way. You do not make a general confusion. You have a clear and comprehensive view of the effects of what you give. Think how you will stand in your farming work if you do things in this way—quite consciously and deliberately. Thus you will gain a knowledge, not for the complication but for the simplification of the fodder problem.

Much—indeed, very much—of what has gradually been discovered by experiment is quite correct. It is only unsystematic, lacking in precision. Precisely this kind of " exact science " is not exact at all in reality, for many things get muddled up together and no one can see through them clearly; whereas the things I have here exemplified can be traced right down into the animal organism in their comparative simplicity, in their comparatively simple mutual effects.

Now take another case. Let us look more towards the flowering nature and the fruiting process that arises in the flower. But we must not stop short at this. We must also observe the fruiting process in the remainder of the plant. Plants have a property which endeared them especially to Goethe. The plant always has throughout its body the inherent potentiality of its specialised parts. For most plants, we put into the earth that which appears as potential fruit in the flower. We plant it in the earth so as to get new plants. With the potato, however, we do not do so. We use the " eyes " of the potato. And so it is in many plants: the fruiting tendency is not only there in the flower. Nature does not carry all her processes to the final stage.

The fruiting process, where Nature has not yet carried it to the final stage, can always be enhanced in its effect by processes which are outwardly similar, in one way or another, to the external process of combustion. For instance, if you chop up and dry the plant for fodder, the stuff you get will be more effective if you let it steam a little—spread it out in the sunlight. The process that is there as an inner tendency is thus led a little farther towards fructification.

There is a wonderful instinct in these matters. Look at the world with intelligence and you will ask: Why did it ever occur to human beings to cook their food? It is a very real question, only as a rule we are not prone to question the everyday things with which we are so familiar. Why did men come to cook their food at all? Because they by and by discovered that a considerable part is played, in all that tends towards the fruiting process, by all such processes as cooking, burning, heating, drying, steaming.

These processes will all of them incline the flower and the seed (yet not only these; indirectly the other parts of the plant also, notably those that lie towards the upper region) to develop more strongly the forces that have to be developed in the metabolic and limb-system of the animal. Even if you take the simple flower or seed—the flower and seed of the plant work on the metabolic or digestive system of the animal. And they work there chiefly by virtue of the forces they unfold, not by their substance. For the metabolic and limb-system requires earthly forces, and in the measure in which it needs them it must receive them.

Think of the animals that pasture on the alpine meadows, for example. They are not like the animals of the plains, for they must walk about under difficult conditions. The conditions are different, simply through the fact that the earth's surface is not level. It is a different thing for animals to walk about on level ground or on a slope. Such animals, therefore, must receive what will develop the forces in the region of their limbs, *i.e.* the forces that have to be exerted by the will. Otherwise they will not become good labouring animals, nor milch-, nor fattening animals.

We must see to it that they get sufficient nourishment from the aromatic alpine herbs, where through the cooking process of the Sun, working towards the flowers, Nature herself has enhanced the fruiting, flowering activity by further treatment. But the necessary force can also be brought into the limbs by artificial treatment, notably if it is anything like cooking, boiling, simmering or the like. Here it is best to take what comes from the fruiting, flowering parts of the plant, and in this way it is especially good to treat such plants as tend from the outset to the fruiting and the flowering—plants, that is to say, which develop little leaf and foliage but tend at once to develop flower and fruit. All that in the plant-world, which does not care to become leaf and foliage, but rather grows rampant in the flowering and fruit-bearing process—that is what we ought to cook.

For themselves, too, men would do well to observe these things. If they did so, we should have less of those movements which take their start from people who find themselves—all unawares—upon the downward slope, the inclined plane of laziness. They say to themselves, no doubt, " If I spend the whole day with petty manipulations, I can never become a true mystic. I can only become a true mystic if I am restful and quiet. I must not always be compelled—by my own needs or by the needs of those around me—to be up and doing. I must be able to say to my surrounding world: I have not the energy to spare for all this outer work. Then I shall be able to become a true mystic. Therefore I will endeavour to arrange my food so that I may become a thorough-going mystic." Well, if you say that to yourself, you will become a raw-food crank. You will have no more cooking. You go in for raw food only.

These things are easily masked; they do not always emerge in this way. If someone who is well on the inclined plane to mysticism of this kind becomes an uncooked-food crank—and if from the outset he has a weak physical constitution—he will make good progress, he will become more and more indolent, *i.e.* mystical.

What happens to man in this respect, we can also apply to the animal. Thus we shall know how to make our animals quick and active. For the human being, however, other things too can occur. He may be physically strong and only afterwards become so " cranky " as to want to be a mystic. He may have strong physical forces in him. Then the processes he has within him—and, moreover, the forces which the raw food itself calls forth in him—will develop strongly, and it cannot do him much harm. For as he eats the raw food he will summon the forces which would otherwise remain latent and create rheumatism and gout. He will summon them to activity, he will develop them and work them and thus grow all the stronger.

Thus there are two sides to every question, and we must realise how all these things are individualised. We cannot give hard and fast principles. This is the real advantage of the vegetarian mode of life. It makes us stronger because we draw forth from the organism forces which would otherwise be lying fallow there. These are, in fact, the very forces that create gout and rheumatism, diabetes and the like.

If we only eat plant food, these forces are called into activity to lift the plant up to human nature. If, on the other hand, we eat animal food from the outset, these forces are left latent in the organism. They remain unused and as a result they will begin to use themselves, depositing metabolic products in various parts of the organism, or driving out of the organs and claiming for themselves things that the human being himself should possess, as in the case of diabetes, etc. We only understand these matters when we look more deeply.

Now let us come to the question, how should we fatten animals? Here we must say to ourselves: As much as possible of cosmic substance must be carried, as it were, into a sack. Oh, the pigs, the fat pigs and sows—what heavenly creatures they are! In their fat body—insofar as it is not nerves-and-senses system—they have nothing but cosmic substance. It is not earthly, it is cosmic substance. The pigs only need the material food they eat, to distribute throughout their body this infinite fulness of cosmic substance which they must absorb from all quarters. The pig must feed, so as to be able to distribute the substance which it draws in from the Cosmos. It must have the necessary forces for the distribution of this cosmic substance.

And so it is with other fattened animals. So you will see: Your fatstock will thrive if you give them fruiting substance (further treated, if possible, by cooking, steaming or the like) and if you give them food which already has the fruiting process in it in a rather enhanced and intensive degree—for instance, turnips or beet, enlarged already in Nature by a process going beyond what they had in them originally—turnips or beet, that is to say, which by enhanced cultivation have grown bigger than they were in the wild.

Once more, then, we can ask ourselves: What must we give to the animal we wish to fatten? Something which will help, at least, to distribute the cosmic substance. It must therefore be something that tends already of its own accord towards the fruiting nature, and that has received the proper treatment in addition. This condition is on the whole fulfilled in certain oil-cakes and the like. But we must not leave the head of such an animal quite unprovided-for. Some earthly substance must still be able to pass upward through this " fattening cure " into the head. We therefore need something else in addition—albeit in smaller quantities, for the head in this instance will not need so much. But in small quantities we do need it. For our fattening animals we should therefore add something of a rooty-nature to the food, however small a dose.

Now there is a kind of substance—indeed, it is pure substance—which has no special function. Generally speaking, we can say, the root-nature has its special functions in relation to the head; the flower in relation to the metabolic and limb-system, and leaf or foliage in relation to the rhythmic system with the substantial nature that belongs to it in the human organism. But there is one more thing whose help we need because it is related to all the members of the animal organisation, and that is the salt-nature. Very little of the food—whether of man or beast—consists of salt! From this *salt-condiment* you can tell that it is not always quantity that matters, but quality. This is important. Even the smallest quantities fulfil their purpose if the quality is right.

Now there is one thing of importance I should like to point out, and I beg you to make exact experiments on this—experiments which could well be extended to an observation of human beings, at any rate of those who incline towards the food question. You know that in modern time (relatively speaking, it is only a short time since) the *tomato* has been introduced as a kind of staple food. Many people are fond of it. Now the tomato is one of the most interesting subjects of study. Much can be learned from the production and consumption of tomatoes. Those who concern themselves a little with these things—and there are such men to-day—rightly consider that the consumption of the tomato by man is of great significance. (And it can well be extended to the animal; it would be quite possible to accustom animals to tomatoes). It is, in fact, of great significance for all *that* in the body, which—while within the organism—tends to fall out of the organism, *i.e.* for that which assumes—once more, within the organism—an organisation of its own.

Two things follow from this. First, it confirms the statement of an American to the effect that a diet of tomatoes will, under given conditions, have a most beneficial effect on a morbid inclination of the liver. In effect, the liver of all organs works with the greatest relative independence in the human body. Therefore, quite generally speaking, liver diseases—those that are rather diseases of the animal liver—can be combatted by means of the tomato.

At this point we can begin to look right into the connection between plant and animal. I may say, in parenthesis, suppose a person is suffering from carcinoma. Carcinoma, from the very outset, makes a certain region independent within the human or animal organism. Hence a carcinoma patient should at once be forbidden to eat tomatoes. Now let us ask ourselves: What is it due to? Why does the tomato work especially on that which is independent within the organism—that which specialises itself out of the organic totality? This is connected with what the tomato needs for its own origin and growth.

The tomato feels happiest if it receives manure as far as possible in the original form in which it was excreted or otherwise separated out of the animal or other organism—manure which has not had much time to get assimilated in Nature—wild manure, so to speak. Take any kind of refuse and throw it together as a disorderly manure- or compost-heap, containing as much as possible in the form in which it just arose—nohow prepared or worked upon. Plant them there, and you will soon see that you get the finest tomatoes. Nay, more, if you use a heap of compost made of the tomato-plant itself—stem, foliage and all—if you let the tomato grow on its own dung, so to speak, it will develop splendidly.

The tomato does not want to go out of itself; it does not want to depart from the realm of strong vitality. It wants to remain therein. It is the most uncompanionable creature in the whole plant-kingdom. It does not want to get anything from outside. Above all, it rejects any manure that has already undergone an inner process. It will not have it. The tomato's power to influence any independent organisation within the human or animal organism is connected with this, its property.

To some extent, in this respect, the *potato* is akin to the tomato. The potato, too, works in a highly independent way, and in this sense: it passes easily right through the digestive process, penetrates into the brain, and makes the brain independent—independent even of the influence of the remaining organs of the body. Indeed, the exaggerated use of potatoes is one of the factors that have made men and animals materialistic since the introduction of potato-cultivation into Europe. We should only eat just enough potatoes to stimulate our brain and head-nature. The eating of potatoes, above all, should not be overdone.

The knowledge of such things will relate agriculture in a most intimate way—and in a thoroughly objective way—to the social life as a whole. It is infinitely important that agriculture should be so related to the social life.

I could go on, giving many individual guiding lines. These guiding lines are only the foundation for manifold experiments, which will extend, no doubt, over a long period of time. Splendid results will emerge if you work out in thorough-going tests and experiments what I have given here. I say this also as a guiding line for your treatment of what has been given in this lecture course.

I am in entire agreement with the strict resolve which has been made by our farmer friends here present, namely, that what has been given here to all those partaking in the Course shall remain for the present within the farmers' circle. They will enhance it and develop it by actual experiments and tests. The farmers' society—the " Experimental Circle " that has been formed—will fix the point of time when in its judgment the tests and experiments are far enough advanced to allow these things to be published.

Full recognition is due to the tolerance which has been shown, which has allowed a number of interested persons, not actually farmers, to share in this Course. They must now recall the well-known opera and fix a padlock on their mouths. Do not fall into the prevalent anthroposophical mistake and straightway proclaim it all from the housetops. We have often been harmed in this way. Persons who have nothing to say out of a real or well-founded impulse, but only repeat what they have heard, go passing things on from mouth to mouth. It has done us much harm. It makes a great difference, for example, whether a farmer speaks of these things, or one who stands remote from farming life. It makes a difference, which you will quickly recognise.

What would result if our non-farmer friends now began to pass these things on, as a fresh and interesting chapter of anthroposophical teaching? The result would be what has occurred with many of our lecture-cycles. Others—including farmers—would begin to hear of these things from this and that quarter. As to the farmers—well, if they hear of these things from a fellow-farmer, they will say, " What a pity he has suddenly gone crazy! " Yes, they may say it the first time and the second time. But eventually—when the farmer sees a really good result, he will not feel a very easy conscience in rejecting it outright.

If, on the other hand, the farmers hear of these things from unauthorised persons—from persons who are merely interested —then indeed " the game is up." If that were to happen, the whole thing would be discredited, its influence would be undermined. Therefore it is most necessary: those of our friends who have only been allowed to take part owing to their general interest and who are not in the Agricultural Circle, must exercise the necessary self-restraint. They must keep it to themselves and not go carrying it in all directions as people are so fond of doing with anthroposophical things.

This principle, as our honoured friend, Count Keyserlingk, to-day announced, has been resolved upon by the Agricultural Circle, and I can only say that I approve it in the very fullest sense. For the rest—except for our final discussion hour—we are now at the end of these lectures. Therefore perhaps I may first express my own satisfaction that you were ready to come here, to take your share in what has been able to be said and in what is now to become of it by further work. On the other hand, I am sure you will all agree with me in this:—

What has here taken place is intended as real, useful work, and as such it has the deepest inner value. But you will bear in mind two things. Let us now think of all the energy and work that was needed on the part of Count and Countess Keyserlingk and all the members of their House to bring to pass all that has come about in this Course. Energy, clear, conscious purpose, anthroposophical good sense, purity and singleness of heart in the cause of Spiritual Science, self-sacrifice and many another thing was necessary to this end. And so it has also come about—I imagine it is so for you all: what we have here been doing as a piece of real hard work, work which is tending to great and fruitful results for all humanity, has been given a truly festive setting by our presence here. We owe it to the way our host and hostess have arranged it all. In five minutes' time you will have another example of their festive hospitality.

All that has been done in this way—last but not least, the cordial kindness of all the people working in the house—has placed our work in the warm and welcome setting of a truly beautiful festival. Thus, with our Agriculture Conference we have also enjoyed a real farm festival. Therefore we offer Countess and Count Keyserlingk and all their House our heartiest and inmost thanks for all that they have done for us in these ten days—for all that they have done in the service of our cause, and for their kind and loving welcome to us all, which has made our sojourn here so pleasant.

DISCUSSION

KOBERWITZ,
16*th June*, 1924.

Question: Has liquid manure the same Ego-organising force as manure itself?

Answer: The essential point is to have the manure and the liquid manure properly combined. Use them in such a way that they work together, each contributing to the organising forces of the soil. The connection with the Ego applies in the fullest sense to the manure, though this does not hold good, generally speaking, for the liquid manure. Every Ego—even the potentiality of an Ego, as it is in the manure—must work in some kind of connection with an astral factor. The manure would have no astrality if " manure-juice " did not accompany it. Thus liquid manure helps—it has the stronger astral force, the dung itself the stronger Ego-force. The dung is like the brain; the liquid manure is like the brain-secretion—the astral force, the fluid portion of the brain, *i.e.* the cerebral fluid.

Question: Might we have the indications as to the proper constellations?

Answer (*by Dr. Vreede*)*:* The exact indications cannot be given now. The necessary calculations cannot be done in a moment. Broadly speaking, the period from the beginning of February until August will hold good for the insect preparations. For field-mice, the periods will vary from year to year. For this year (1924) the time from the second half of November to the first half of December would be right.

Dr. Steiner: The principles of an anthroposophical calendar, such as was planned at the time, should be carried out more fully. Then you could follow such a calendar precisely.

Question: Speaking of full Moon and new Moon, do you mean the actual day of the full or new Moon, or do you include the time shortly before and after?

Answer: You call it new Moon from the moment when this picture appears, approximately speaking (Diagram 22). This picture is there; then it vanishes. And you reckon it full Moon from the time when the following picture occurs. New Moon, therefore, from the time when the Moon appears as a quite narrow crescent, and then disappears. Twelve to fourteen days in each case.

Question: Can insects, unobtainable at the season of the given constellation, be kept until the proper time arrives?

Answer: We shall give more exact indications of the time when the preparations should be made. The several forms of insects can no doubt be kept.

Question: Must the weed-seeds be burnt in summer, or can it be done at any time?

Answer: Not too long after collecting the seed.

Question: What of the sprinkling of insect-pepper taken from insects that have never come into actual contact with the earth?

Answer: Sprinkle it on the earth just the same. For the insect, the process does not depend on physical contact, but on the quality communicated by these homoeopathic doses. The insect has quite another kind of sensitiveness; it flees from what ensues when the preparation is sprinkled in the earth. That the insect does not come into direct contact with the earth makes no difference at all.

Question: What of the harmfulness of frost in farming, especially for the tomato? In what cosmic relationship is frost to be understood?

Answer: If the tomato is to grow nice and big, it must be kept warm; it suffers greatly from frost.

As to frost in general, you must realise what it is that comes to expression in the effects of frost. These effects always represent a great enhancement of the cosmic influences at work in the earth. This cosmic influence has its normal mean when certain degrees of temperature are prevalent; then it is just as the plant requires it. If, on occasion, we get frost of long duration or too intense and deeply penetrating, the influence of the heavens on the earth is too strong, and the plants will tend to ramify in various directions, to form thread-like growths, to spread out thinly. And the resulting growths, being thin, will under certain conditions naturally be received by the prevailing frost, and destroyed. Frost, therefore, when it goes too far, is undoubtedly harmful to plant-growth, simply because too much of the heavens comes into the soil of the earth.

Question: Should one treat the bodies of animals with the burnt relics of horse-flies and the like, or should these relics be scattered over the meadows and pastures?

Answer: Wherever the animal feeds. Sprinkle the relics over the fields; they are all to be thought of as additions to the manure.

Question: What is the best way of combatting couch-grass? It is very difficult, is it not, to get the seeds?

Answer: The mode of propagation of the couch-grass you have in mind—where it never goes so far as to form seed—will in the end eliminate itself. If you get no seed, you have not really got the weed. If, on the other hand, it establishes itself so strongly that it plants itself and continues to grow rampantly, you then have the means to combat it, for you will soon find as much seed as you require, because, in fact, you need so very little. After all, you can also find four-leaved clover.

Question: Is it permissible to conserve masses of fodder with the electric current?

Answer: What would you attain by so doing? You must consider the whole part played by electricity in Nature. It is at least comforting that voices are now being heard in America—where, on the whole, a better gift of observation is appearing than in Europe—voices, I mean, to the effect that human beings cannot go on developing in the same way in an atmosphere permeated on all sides by electric currents and radiations. It has an influence on the whole development of man.

This is quite true; man's inner life will become different if these things are carried as far as is now intended. It makes a difference whether you simply supply a certain district with steam-engines or electrify the railway lines. Steam works more consciously, whereas electricity has an appallingly unconscious influence; people simply do not know where certain things are coming from. Without a doubt, there is a trend of evolution in the following direction. Consider how electricity is now being used above the earth as radiant and as conducted electricity, to carry the news as quickly as possible from one place to another. This life of men in the midst of electricity, notably radiant electricity, will presently affect them in such a way that they will no longer be able to understand the news which they receive so rapidly. The effect is to damp down their intelligence. Such effects are already to be seen to-day. Even to-day you can notice how people understand the things that come to them with far greater difficulty than they did a few decades ago. It is comforting that from America, at least, a certain perception of these facts is at last beginning to arise.

It is a remarkable fact that whenever something new appears, as a rule in the early stages it is heralded as a remedy—a means of healing. Then the prophets get hold of it. It is strange, where a new thing appears, clairvoyant perception is often reduced to a very human level! Here is a man who makes all sorts of prophecies about the healing powers of electricity, where no such thing would previously have occurred to him. Things become fashionable! No one was able to imagine healing people by electricity so long as electricity was not there. Now—not because it is there, but because it has become the fashion—now it is suddenly proclaimed as a means of healing. Electricity—applied as radiant electricity—is often no more a means of healing than it would be to take tiny little needles and prick the patient all over with them. It is not the electricity—it is the shock that has the healing effect.

Now you must not forget that electricity always works on the higher organisation, the head-organisation both of man and animal; and correspondingly, on the root-organisation in the plant. It works very strongly there. If, therefore, you use electricity in this way—if you pour electricity through the foodstuffs—you create foodstuffs which will gradually cause the animal that feeds on them to grow sclerotic. It is a slow process; it will not be observed at once. The first thing will be, that in one way or another the animals will die sooner than they should. Electricity will not at first be recognised as the cause; it will be ascribed to all manner of other things.

Electricity, once for all, is not intended to work into the realm of the living—it is not meant to help living things especially; it cannot do so. You must know that electricity is at a lower level than that of living things. Whatever is alive—the higher it is, the more it will tend to ward off electricity. It is a definite repulsion. If now you train a living thing to use its means of defence where there is nothing for it to ward off, the living creature will thereby become nervous or fidgetty, and eventually sclerotic.

Question: What does Spiritual Science say to the preservation of foodstuffs by acidification, as in the silage-process?

Answer: If you are using salt-like materials at all in the process —taken in the wider sense—it makes comparatively little difference whether you add the salt at the moment of consumption or add it to the fodder. If you have fodder with insufficient salt-content to drive the foodstuffs to the parts of the organism where they should be working, the souring of such fodder will certainly be beneficial.

For instance, suppose you have turnips, swedes, etc., in a certain district. We have seen that they are especially fitted to influence the head-organisation. They are excellent fodder for certain animals—young cattle, for example. If, on the other hand, in some district you notice that as a result of such fodder the animal tends to lose hair too early or too much, then you will salt the fodder. For you will know that it is not being sufficiently deposited at those parts of the organism which it should reach; it is not getting far enough. Salt, as a rule, has an exceedingly strong influence in this direction, causing a foodstuff to reach the place in the organism where it ought to work.

Question: What is the attitude of Spiritual Science to the ensiling of the leaves of sugar-beet, etc., and other green plants?

Answer: You should see that you get the optimum effect; you must not go beyond the optimum in the method used. Generally speaking, the souring will not have a harmful effect unless carried to excess by the addition of excessive quantities of admixtures. For the salt-like constituents are precisely those that tend most strongly to remain as they are in the living organism.

Usually the organism (the animal organism also, and the human to a still greater extent) is so constituted that it changes whatever it absorbs in the most manifold ways. It is mere prejudice to think, for example, that any part of the protein you introduce through the stomach is still available after this point in the same form in which you introduce it. The protein must be completely transformed into dead substance, and must then be changed back again by the etheric body of man himself (or of the animal) into a protein which is then specifically human or animal protein.

Thus, everything that penetrates into the organism must undergo a complete change. What I am saying applies even to the ordinary warmth. I will draw it diagrammatically (Diagram 23). Assume that you have here a living organism; here you have warmth in its environment. Suppose on the other hand that you here have a piece of wood, which, though it comes from a living organism, is already dead, and you have warmth in its environment. Into the living organism the warmth cannot simply penetrate; it does not merely penetrate it. The moment the warmth begins to come inside, it is already worked upon by the living organism; it changes into warmth that has been assimilated and transmuted by the living organism itself. Indeed, it cannot rightly be otherwise. Into the dead wood, on the other hand, the warmth will simply penetrate; the warmth inside is the same as in the surrounding mineral kingdom of the earth.

Not so with living bodies. The moment any warmth begins to penetrate unchanged into our organism, for example—as it would penetrate into a piece of wood—that moment, we catch cold. Whatever enters from outside into the living organism must not remain as it is; it must at once be changed. This process takes place least of all in salt. Hence, with the salts, used in the way you indicate for ensiling the foodstuffs—provided you are just a little sensible and do not give too much (for then in any case the animal would reject the food because of its taste)—you will do no great harm. If it is necessary for preservation, that in itself is a sign that the process is right.

Question: Is it advisable to ensile the fodder without salt?

Answer: That is a process much too far advanced. It is, I would say, a super-organic process. When it has gone too far, it can under certain circumstances be extremely harmful.

Question: Is the Spanish whiting (sometimes used to mitigate the souring effects) harmful to animals?

Answer: Certain animals cannot stand it at all; they become ill at once. Some animals can stand it; I cannot say which at the moment. Generally speaking, it will not do the animals much good; they will tend to become ill.

Question: I imagine the gastric juice will be dulled by using it?

Answer: Yes, it will be made ineffective.

Question: I should like to ask if it is not of great importance in what frame of mind one approaches these matters? It makes a great difference whether you are sowing corn or scattering a preparation for destructive ends. Surely the attitude of mind must come into question. If you work against the insects by such means as are here indicated, will it not have a greater karmic effect than if in single instances you get rid of the animals by some mechanical means?

Answer: As to the attitude of mind—surely the chief point is whether it be good or bad! What do you mean by the " destruction "? You need but consider the whole way in which you have to think about these things in any case. Take to-day's lecture, for instance, and the way it has been held; when, for example, I pointed out how one must know about the things of Nature: how one must see from the outer appearance, say, of the linseed or the carrot, what kind of process it will undergo inside the animal.

You will go through such an objective education if this knowledge becomes a reality in you at all, that it is surely quite unthinkable without your being permeated with a certain piety and reverence. Then you will also have the impulse to do these things in the service of mankind and of the Universe.

If harm were to result from the spirit in which you do them, it could only be a question of your bringing in deliberately evil intentions. Yes—you would have to have downright bad intentions. If, therefore, common morality is at the same time fostered, I cannot imagine how it should have bad effects in any way. Do you conceive that to run after an animal and kill it would be less bad?

Question: I was referring to the manner of destruction—whether it be by mechanical means, or by these cosmic workings—whether that makes a difference.

Answer: This question raises very complicated issues, the understanding of which depends upon your seeing them in large connections. Let us assume, for instance, that you draw a fish out of the sea and kill it. Then you have killed a living thing. You have carried out a process which takes place upon a certain level. Now let us assume that for some purpose you scoop up a vessel of sea-water in which much fish-spawn is contained. You will thus be destroying a whole host of life. Thereby you will have done something very different than in destroying the single fish. You will have carried out a process on an entirely different level.

When such an entity in Nature passes on into the finished fish, it has followed a certain path. If you reverse this path, you are bringing something into disorder. But if I hold up, at an earlier stage, a process which is not yet completed (or which has not yet come to an end in the blind-alley of the finished organism), then I have not by any means done the same thing as when I kill the finished organism.

I must therefore reduce your question to this: What is the wrong I do when I make the pepper? What I destroy by the pepper scarcely comes into question. The only thing that could come into question would be the creatures I need to *make* the pepper. And to do this, I shall obviously in most cases destroy far fewer animals than if I had to catch them all with much trouble, and kill them. I fancy, if you think it over in a practical way and not so abstractly, it will no longer seem to you so monstrous.

Question: Can human faeces be used, and to what treatment must it be submitted before use?

Answer: Human faeces should be used as little as possible. It has very little effect as manure, and it is far more harmful than any kind of manure could possibly be. If you *will* use human faeces, so much as will find its way into the manure of its own accord on a normal farm is quite sufficient. Take that as your maximum measure of what is not yet harmful. You know there are so and so many people on a normal farm, and if with all the manure you get from the animals and in other ways there is also mixed what comes from the human beings—that is the maximum amount which may be used.

It is the greatest abuse when human manure is used in the neighbourhood of large cities; for in large cities there is enough for an agricultural district of immense proportions. Surely you cannot fall a prey to the demented idea of using up the human dung on a small territory in the neighbourhood of a large city—say, Berlin. You need only eat the plants that grow there; they will soon show you what it means. If you do it with asparagus, or anything that remains more or less sincere and upright, you will soon see what happens.

Moreover, you must bear in mind that if you eat this kind of dung for growing plants which animals will eat, the eventual result is even more harmful, for in the animals much of it will remain at this level. In passing through the organism, many things remain at the level which the asparagus preserves when it goes through the human body. In this respect crass ignorance is responsible for the most awful abuses.

Question: How can red murrain (Erysipelas) in swine be combatted?

Answer: That is a veterinary question. I have not considered it, because no one has yet asked my advice about it. But I think you will be able to treat it by external applications of grey antimony ore in the proper doses. It is a veterinary, a medical question, for this is a specific disease.

Question: Can the Wild Radish*, which is a bastard, also be combatted with these peppers?

Answer: The powders of which I have spoken are specifically effective only for the plants from which they are derived. Thus, if a plant is really the outcome of crossing with other species, one would expect it to be immune. Symbioses will not be affected.

Question: What about green manuring?

Answer: It also has its good side, especially if you use it for fruit-culture, in orchardry. Such questions cannot be answered in an absolutely general way. For certain things, green manuring is useful. You must apply it to those plants where you wish to induce a strong effect on the growth of the green leaves. If this is your intention, you may well supplement other manures with a little green manuring.

* *Raphanus raphinastrum.*

SUPPLEMENT

The following is a record of indications given verbally by Dr. Steiner to individuals in answer to questions and with reference to particular problems and local conditions. (Several of these were given prior to the Agriculture Course of June, 1924.)

Readers should remember that they are quoted from memory, are fragmentary and not necessarily of universal application.

SUPPLEMENT

The following indication was given by Dr. Steiner at the Guldesmühle Mill in Dischingen during a conversation about the more or less harmful influences of artificial mineral manures. Dr. Steiner said that in view of the increase in yield which was generally required, they might perhaps not be able to forego the use of such manures. But the harmful influence, for human beings and for animals alike, would not fail to ensue. Some of these influences would not appear in full till generations after. At any rate it was necessary to discover and apply remedial measures in good time. Such, for example, were the leaves of fruit-trees, and it was therefore good to plant fruit-trees on the fields.

A second indication by Dr. Steiner concerned the use of horn manure. This had been manufactured at the Guldesmühle Mill, and it was further developed at Einsingen. In answer to a direct question as to the value of horn manure, Dr. Steiner replied that mixed with ordinary stable manure, horn manure was among the very best. Subsequently we asked Dr. Steiner whether roasted or unroasted horn-meal was better. (At Einsingen we do not roast it, whereas as a general rule the horn-shavings, etc., are first subjected to a very rigorous drying process. The advantage is that they are more easily ground down after this process. On the other hand, the roasting involved a loss of about 15 per cent, consisting mainly of water). Dr. Steiner answered to the effect that unroasted horn-meal was better on account of the higher hydrogen content. For the right influence of the manure, the hydrogen content was in fact far more important even than the nitrogen, though modern science had not yet awakened to the real importance of the hydrogen content for plant growth.

—*Communicated by Dr. Rudolf Maier.*

REPORT OF A CONVERSATION BETWEEN DR. STEINER AND DR. STREICHER

Dr. Streicher: Another matter we are concerned with here is one that was brought very near to me in my youth. I grew up in the country, and was much concerned with the problem of manures for plant-life generally. The present position—the prevalent opinion on these matters—seems to me highly detrimental. The prevailing notions about manures have not gone far beyond what was inaugurated by Liebig, who wanted to instil mineral substances into the soil—nitrogen, phosphoric acid and potassium, for instance. The artificial manure industry in its present stage produces nitrogen bound to very strong acids—hydrochloric and sulphuric. Agriculture is faced with a new danger, which has even now become reality to some extent. Artificial manures are brought into the soil, regardless of the way the plants receive them. These artificials give rise to an acid reaction in the soil, and in a dry summer the results are disastrous.

Dr. Steiner: The fact is, the only really sound manure is cattle-manure. The first principle is to take one's start from this. It is the really healthy manure. At the same time, a healthy nitrogen content must be brought about in the soil by discovering some principle, by virtue of which the soil will be thoroughly worked-through by earth-worms and similar creatures. I do not think we have yet gone so far as to be able to tell quite fully what this is.

Then it will also be essential to find the necessary weeds—in a word, the necessary neighbour-plants. As I said yesterday to Herr St——, who is now devoting himself to Agriculture, it is important, for example, to plant sainfoin on the rye- and wheat-fields, at least along the edges. This influence decidedly exists. You should investigate scientifically how important it is to plant horse-radish along the edge of your potato fields, to have a sprinkling of cornflowers in your corn fields, and to exterminate the poppy.

These things should be considered in connection with the manuring question as a whole. Otherwise you are reduced to the most abstract principles, where for example you get acids formed in the soil, and you then ask: " How can I counteract them?" and on these lines, in course of time, you absolutely kill the soil for plant-growth. You make it deaf.

Dr. Streicher: The farmers too have a feeling that the soil is extracted and impoverished by the use of artificial manures.

Dr. Steiner: It is not at all a bad expression; it makes the soil deaf. On the other hand, one must not fall into the extreme of using plant-manure. It must be admitted that plant-manure is not favourable to plant-growth. In point of fact, the only ideal manure is cattle-manure—not plant-manure. Everything follows on this basic principle. Also you must be clear that very much depends on the neighbouring plants, notably leguminosae—sainfoin especially. With herbaceous plants you should also take care as far as possible to plant them in a dry soil, whereas with cereals a moist soil is needed.

Moreover, strange as it may sound to the chemist and biologist of to-day, your human and personal relation to the seed-corn is undoubtedly important. If you examine it thoroughly, you will find it makes a difference to the thriving of the corn, whether the sower simply takes the seed-corn out of a sack and throws it down roughly, or whether he has the habit of shaking it a little in his hand and throwing it gently, sprinkling it on the ground. These differences are of importance in relation to the manuring problem.

It would be good for you to discuss these matters with farmers, who cannot but be interested in them. They have no little experience, only their experiences are eclipsed nowadays. Modern agriculture has such experience no longer. Altogether I should advise you—I think it will serve you well—to use old peasant-calendars in connection with manuring problems. They contain very curious instructions, some of which you will indeed be able to formulate in chemical terms.

Dr. Streicher: It is difficult for the modern farmer, especially just now. Last year the stock of cattle was much reduced by illness; and it has very largely been reduced by lack of fodder.

Dr. Steiner: Scientists will have to summon up courage to point out the main detrimental causes. The undue praise of stable feeding in recent times is undoubtedly connected with the prevalent tuberculosis among cattle. For all I know, the animals may be able to give more milk for a short time, or what not; but their state of health deteriorates through generation after generation. It should go without saying.

Even the manure which the peasant-woman—basket on back and shovel in hand—gleans from the meadows, is undoubtedly better than the manure you get by stable-feeding. Also the animals ought not to have to absorb the breath of the neighbouring animal while they are feeding; that is undoubtedly harmful.

Go out on to the pastures and you will see, they keep a certain distance apart. Look at the pastures for once, and you will find that of their own accord the beasts take their stand at a considerable distance from one another. The animal cannot abide the breath of the neighbouring animal while it is feeding. And, after all, how easily it occurs that an animal gets an abrasion, and if the breath of the neighbouring beast comes into this, it will undoubtedly be a cause of disease.

Dr. Streicher: Perhaps I may point out certain prevailing tendencies in outer science—in the use of artificial manures and synthetic materials? Having succeeded in the synthetic fabrification of nitrogen products, they are now boasting the discovery of the synthesis of protein. They find it tedious to have to go *via* the plants in gaining protein. There is already a movement on foot to short-circuit this " roundabout way " of the plant, and to feed the animals on synthetic nitrogen manure directly.

It may sound strange, but scientists have made investigations on these lines. They set great store by the synthetic urea which is added as a concentrated foodstuff to the ordinary hay, as cattle fodder. It has also been tried on sheep. The idea is that certain bacteria live in the paunch of the animal, and that these bacteria will disintegrate the urea and transform it into albumen or protein. I think the danger is very real. If these experiments are continued—if it becomes habitual among farmers to give urea and other synthetic foods—the present symptoms of deterioration in our stock will go from bad to worse.

Dr. Steiner: True results can never follow from experiments conducted in this way. In the sphere of vitality—if I may so express it—there is always the law of inertia. That is to say, it may not appear in the present generation or in the next, but it will in the third. The vitalising influence goes on beyond the first few generations. If you restrict your investigations to the present and do not extend them over several generations, you get a completely false picture. Then, when you do observe the next generation but one, you turn your attention to quite other causes than the real ones, namely, the feeding of the grandparent beasts. Vitality cannot be broken down at once. It is surely broken, but only in succeeding generations.

Dr. Streicher: In studying this question last year, I came upon a piece of work that gained publicity in England during the war—I mean the researches of the English botanist, *Bottomley*. Bottomley discovered that there are certain plants which cannot absorb mineral manure directly. If you make a solution of nutritive salts, certain plants cannot live in it for long. On the other hand, he observed that if a certain bacterial life was brought about in the soil, substances were thereby formed which he could not quite get hold of chemically. He puts them side by side with the " vitamins " of the biologists. Adding these substances in imponderable quantities to the nutritive salt solution, he finds that the plants unfold a quite extraordinary life. The substances he thus produces he describes as " auxines "—life-kindling substances. During the war, when England was obliged to till the soil for the growth of cereals, this " Humogen "—as it was named by Bottomley—was produced in large quantities and added to the earth. In certain cases it had an extraordinary effect; in other cases the effect was absent.

Dr. Steiner: Which plants received this blessing?

Dr. Streicher: It is not said.

Dr. Steiner: Food-plants?

Dr. Streicher: In the growth of cereals. . . .

Dr. Steiner: If it is done with food-plants, the people who consume them will suffer no great harm, but their children may very well be born with hydrocephalus. From the whole process it is evident that the development of the plant has been hypertrophied. When such plants are used for nourishment, the result is a malformation of the nervous life in the next generation. This is the fundamental fact: certain effects in the life-process only shew themselves in the next generation, or even only in the next but one. So far must the investigations be extended.

Dr. Streicher: One could mention in the same connection the experiments initiated by a Freiburg scientist. He made organic mercury salts and manured the vegetable gardens with them during the war. Growth was remarkably enhanced by this " mercury-manuring." People even began to hope that the whole question of plant-growth would rapidly be solved; that vegetables would be produced in a very short time. These vegetables too shewed a hypertrophied growth.

Dr. Steiner: You would have to investigate whether the children of those who consume them do not grow up impotent. These things must all be examined, for in this sphere you simply cannot make your experiments within narrow limits. The vital process goes on in time, and only in the course of time does it degenerate in its inherent forces.

FURTHER INDICATIONS BY DR. STEINER RELATING TO AGRICULTURE

Dr. Steiner gave the following answers to questions by *Herr Stegemann:—*

In preparing the ground for oats, one should take care that the soil is dry. So, too, for potatoes and root-crops. Wheat and rye on the other hand should be sown in a moist soil.

As border-plants for cereals, Dr. Steiner indicated dead-nettle and sainfoin. They should be planted four to five metres apart. Horse-radish might be good as a border-plant for roots and potatoes. It need only be planted at the four corners of the plot. It must be eradicated every year.

Concerning animal pests, Dr. Steiner remarked that as new cultivated plants were evolved, they would increasingly disappear.

Against wire-worm, Dr. Steiner gave the following method: Expose rain-water to the waning moon for a fortnight, and then pour the water over the places where the worm occurs. One should take enough water to moisten the soil through to the level where the worm abides.

To counteract the deterioration of the potato, Dr. Steiner said the seed-potato should be cut into pieces until every little piece has only a single eye. The same process should be repeated in the following year.

* * *

In answer to questions by *Count Carl von Keyserlingk*, *Dr. Steiner* gave the following indications (*communicated by Count Adalbert Keyserlingk*)*:—*

To counteract smut, a ring of stinging-nettles should be planted round the fields. On the same occasion, Dr. Steiner remarked that it is good to put the manure-heaps on the field until the time when the manure is needed. For an orchardry on a rather moist and boggy soil, Dr. Steiner recommended " Kali magnesia."

When walking through the flower gardens at Whitsun, 1924, Dr. Steiner remarked as he looked at the flowers: " They none of them seem to feel quite happy here; there is too much iron in the soil." When he came to the roses, which were not flowering well, and did not look at all healthy (mildew), Dr. Steiner advised that very finely divided lead be given to the soil.

When it was pointed out that an enormous number of cow-horns would surely be needed for the Koberwitz estate—an area of

18,500 acres—Dr. Steiner gave the astonishing reply that once it was all in working order, probably no more than 150 cow-horns would be needed for this land.

To a question by *Count Wolfgang von Keyserlingk* on the use of sainfoin, Dr. Steiner answered that abour 2 lb. of sainfoin seed should be included with the seed-corn per three-fifths acre.

* * *

Question: In Dornach and Arlesheim we suffer from an awful plague of slugs. They eat up all the foliage.

To counteract them, Dr. Steiner advised the following remedy: Sprinkle out a 3-in-1,000 dilution of pine-cone seeds. The answer is to be understood as follows: The soluble content of the seeds (which must presumably be extracted by pressure) should be dissolved in water to a dilution of 3-in-1,000, and this should then be sprinkled over the beds affected. Dr. Steiner said we should begin by making this experiment. It would be very interesting if parallel experiments were made at other places.

Once when we were going round the Dornach and Arlesheim plantations, Dr. Steiner advised the following method of strengthening preparation " 500 " for the meadow-land—for the land where fruit-trees were standing. Take a few fruits and a handful of leaves of the kind of fruit in question; make a decoction of these with a litre of water, and add this fruit-decoction to the bucket in which the content of the horn is being stirred.

To strengthen sick and feeble fruit-trees, a circular trench about a hand's-breadth deep may be dug around the tree in a circumference approximately corresponding to the crown of the tree. Into this trench pour larger quantities of the stirred-up preparation " 500."

For the silica preparation " 501," Dr. Steiner said it would even suffice to mingle and knead up a piece of quartz of the size of a bean with soil from the land which is afterwards to be sprinkled, and put this mixture into the horn. This would already contain sufficient silica-radiation if a little of it was dissolved and stirred.

As border plants for vegetable gardens, sainfoin, dandelion and horse-radish were mentioned.

To a question about plant-diseases, Dr. Steiner answered: Properly speaking, there can be no such thing as sick plants, for the etheric is always healthy. If disturbances occur in spite of this, it is a sign that something is wrong with the environment of the plant, especially the soil. To strengthen trees that are growing old, he said we might try the effect of putting fresh earth around their roots—earth taken from the neighbourhood of the roots of sloe (*Prunus spinosa*) and birch.

To make the destruction of weeds more effective, the root-stock *and* seed of the weed may be burned.

Communicated by Ehrenfried Pfeiffer.

Some years before the War, Dr. Steiner said, in answer to a question about the use of night-soil: It should not be used at all, because the cycle from man to plant and back again to man is too short. (The question referred to gardening.) The proper cycle is from man to plant, from plant to animal, from animal to plant ; then only from the plant again to man.

Dr. Steiner repeatedly and expressly rejected the use of peat for the improvement of the soil, whether as manure or as a would-be improvement of the physical properties of the soil. Humus and humus again should be given to the soil in every conceivable form —as compost, leaf-mould, etc.

Communicated by Frl. Gertrud Michels.

* * *

To a question on the use of mineral manure (compare page 70 of the Course), Dr. Steiner answered: If obliged to use mineral manure, one should always mix it first with dung or liquid manure. Dr. Steiner strongly rejected the use of lavatory fluid. It should not even be emptied out on to fresh compost—" not even if the compost-earth will only be needed after four years. Even then, things are contained in it which are not good."

Communicated by Frau A. Ganz.

* * *

Under trees that suffer from woolly aphis (*Eriosoma lanigerum*), a ring of nasturtiums should be planted.

Communicated by Franz Lippert.

INDEX

RUDOLF STEINER (1861–1925) called his spiritual philosophy 'anthroposophy', meaning 'wisdom of the human being'. As a highly developed seer, he based his work on direct knowledge and perception of spiritual dimensions. He initiated a modern and universal 'science of spirit', accessible to anyone willing to exercise clear and unprejudiced thinking.

From his spiritual investigations Steiner provided suggestions for the renewal of many activities, including education (both general and special), agriculture, medicine, economics, architecture, science, philosophy, religion and the arts. Today there are thousands of schools, clinics, farms and other organizations involved in practical work based on his principles. His many published works feature his research into the spiritual nature of the human being, the evolution of the world and humanity, and methods of personal development. Steiner wrote some 30 books and delivered over 6000 lectures across Europe. In 1924 he founded the General Anthroposophical Society, which today has branches throughout the world.

Lecture	Date	1958 Edition	1929 Translation
1	7th	17- 28	I/ 1-13
2	10th	29- 41	II/ 1-16
3	11th	42- 56	III/ 1-18
Address	11th	57- 64	D/ 1-11
4	12th	65- 76	IV/ 1-15
D.	12th	77- 86	D/12-24
5	13th	87-100	V/ 1-19
D.	13th	101-106	D/25-32
6	14th	107-119	VI/ 1-17
D.	14th	120-124	D/33-39
7	15th	125-135	VII/ 1-13
8	16th	136-151	VIII/ 1-20
D.	16th	152-158	D/40-49
Supplement		159-167	A/ 1-11
Index		169	